ISLAM & EVOLUTION SERIES

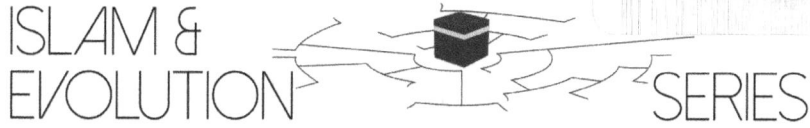

Islam and Evolution is a groundbreaking series that speaks openly and honestly about evolution from a wide range of perspectives. Aimed at general readers, it brings together scholars and academics from diverse intellectual, cultural, and disciplinary backgrounds to explore the complex and often sensitive relationship between Islam and evolutionary theory.

What sets this series apart is its commitment to intellectual openness: it neither promotes a fixed doctrinal stance nor avoids controversy. Instead, it offers a welcoming space for multiple interpretations, fostering thoughtful reflection and genuine dialogue. Whether readers come with questions, convictions, or curiosity, Islam and Evolution encourages a deeper and more nuanced engagement with one of the most significant conversations in science and religion today.

ABOUT THE AUTHORS

Rui Diogo is a multi award-winning researcher, speaker and writer. Worldwide, he is renowned for addressing broader scientific questions and societal issues using empirical data from many different fields of science. He obtained his Bachelor's degree in Biology from the University of Aveiro, Portugal, and a PhD at the University of Liege, Belgium. This was followed up with a postdoc. at King's College London, and a Masters and PhD at the Department of Biology at George Washington University, US.

A wanderer as well as a wonderer, he has conducted research, given speeches, and travelled to more than 120 countries. He is the author of more than 150 papers in top journals and 20 books, including: "*Learning and Understanding Human Anatomy and Pathology,*" – a textbook that is used in several medical schools globally – "*Evolution Driven by Organismal Behavior*" – listed amongst the ten best evolutionary books of 2017 – and the highly acclaimed "*Meaning of Life, Human Nature, and Delusions.*"

Aamina Malik is a medical student at Howard University College of Medicine in Washington, D.C. She earned a Bachelor's degree in Biology and a Masters degree in Anatomy – under the supervision of Rui Diogo – from Howard University in 2011 and 2015, respectively. She earned her medical degree in May 2021.

Malak Alghamdi earned her Bachelor's degree in Physical Therapy from King Abdulaziz University in Jeddah, Saudi Arabia. She followed this with a Masters and PhD in Anatomy from Howard University in 2015 and 2018, respectively – under the supervision of Rui Diogo. Currently, she is an Assistant Professor at King Saud bin Abdulaziz University for Health Sciences in Jeddah, Saudi Arabia.

A Thousand Years Before Darwin

*Untold Stories of Muslim Scholars' Contributions to Anatomy
and Evolutionary Biology*

A Thousand Years Before Darwin
Untold Stories of Muslim Scholars' Contributions to Anatomy and Evolutionary Biology

Rui Diogo, Aamina Malik, Malak Alghamdi

First published in the UK by Beacon Books and Media Ltd
Earl Business Centre, Dowry Street, Oldham OL8 2PF UK .

Copyright © Rui Diogo 2025

The right of Rui Diogo to be identified as the author of this work has been asserted in accordance with the Copyright, Designs and Patents Act 1988. All rights reserved. This book may not be reproduced, scanned, transmitted or distributed in any printed or electronic form or by any means without the prior written permission from the copyright owners, except in the case of brief quotations embedded in critical reviews and other non-commercial uses permitted by copyright law.

First edition published in 2025

www.beaconbooks.net

ISBN 978-1-915025-24-1 Paperback
ISBN 978-1-915025-25-8 Hardback
ISBN 978-1-915025-26-5 Ebook

Cataloging-in-Publication record for this book is available from the British Library

Cover design by Raees Mahmood Khan

Book Dedication

"The moment a little boy is concerned with which is a jay and which is a sparrow, he can no longer see the birds or hear them sing."
(Eric Berne)

This book is dedicated to all humans, from any society, group, religious background, or geological time, who have allowed us—with their ideas, discussions, or scientific discoveries—to better understand the reality of the natural world.

CONTENTS

ACKNOWLEDGEMENTS xi

PREFACE xiii

1. BIASES IN SCIENCE AND HISTORY OF SCIENCES 1
   Biased Reactions to Untold Stories About Discoveries and Ideas of
   Muslim Scholars 1
   Darwin's Own Beliefs, Biases and Prejudices 6
   Darwin's Mother Nature, Her 'Broom' and the Notion of 'Progress' 15
   Other Examples of Political and Societal Influences on Science and the
   History of Science 22

2. UNTOLD STORIES: MUSLIM SCHOLARS IN
   ANATOMICAL AND MEDICAL SCIENCES 35
   Seeing Human Anatomy in the Body of Macaques 35
   Major pre-Vesalius Muslim Scholars who Studied Human Anatomy 48
   - Al-Razi, Abu-Bakr Muhammad ibn Zakariya (Rhazes)
     865–925, Al-Rayy, Iran (Persian) 49
   - Al-Akhawayni Bukhari, Abu Bakr Rabi ibn Ahmad (Joveini)
     ?–983, Bukhara, Republic of Uzbekistan (Persian) 54
   - Ibn Abbas, Abu al-Qasim Ismail ibn Abbad ibn al-Abbas ibn Abbad
     ibn Ahmad ibn Idris (Haly/Hali Abbas)
     930/949–994, Arejan, Iran (Persian) 56
   - Ibn Sina, Abu Ali Husain ibn Abdullah (Avicenna)
     980–1037, Afshaneh, near Bukhara (Persian) 74
   - Ibn Al-Haytham, Abu Ali Al-Hasan ibn Al- Hasan (Al-Hazen)
     965–1040, Basra, Iraq (Arab) 92
   - Ibn Rushd, Abu Al-Walid Muhammad ibn Ahmad ibn Muḥammad
     (Averroes)
     1126–1198, Cordoba, Spain (Arab) 95
   - Al-Baghdadi, Muwaffaq Al-Din Muhammad Abd Al-Latif ibn
     Yusuf
     1162–1231, Baghdad, Iraq (Arab) 99

- Ibn Al-Nafis, Abu Al-Hasan, Ala'a Al-Din Ali ibn Abi Al-Hazm Al-Qurashi Al-Dimashqi
  1210–1288, Damascus, Syria (Arab) 101
- Mansur ibn Ilyas, Mansur ibn Mohammad ibn Ahmad ibn Yousef ibn Ilyas
  1380–1409, Shiraz, Timurid Persia 106

General remarks on the anatomical discoveries of pre-Vesalius Muslim scholars 110

3. **UNTOLD STORIES: MUSLIM SCHOLARS AND EVOLUTIONARY IDEAS** 113

   Religion, creationism, human evolution, and stereotypes 113
   - Al-Jahiz 124
   - Ibn Miskawayh 127
   - The Ikhwan Al-Safa 129
   - Al-Beruni 132
   - Ibn Tufayl 134
   - Nidhami Arudi 137
   - Tusi 139
   - Ibn Khaldun 141

   General remarks about the transmutation ideas of Muslim scholars 142

4. **BRINGING REALITY TO SCIENCE AND SOCIETY** 145

REFERENCES AND SUGGESTED FURTHER READING 153

FIGURE CREDITS 179

INDEX 181

Acknowledgements

"Raise your words, not voice. It is rain that grows flowers, not thunder."
(Rumi)

We would like to thank Jason Wiles, Mohammadali Shoja, Shane Tubbs, Tatjana Buklijas, Hakima Amri, Martin Chalfie, Monica Green, Nahyan Fancy, Fatimah Jackson, Cecilia Veracini and Maud Kozodoy for their comments regarding the history of anatomy, biology, sciences, and/or the Muslim world. Whether minor or major, these remarks were very helpful for our projects on the contributions of Muslim scholars to the history of anatomical knowledge and evolutionary thought.

We are also thankful to the editors of the journals who published the two papers which are the main basis of this book, and, in particular, the former editors of *Anatomical Record*, Jeffrey Laitman and Kurt Albertine, for their wise and valuable comments, and for featuring one of these papers on the cover of the issue. This prominence paved the way for the worldwide dissemination of this paper.

Among the many journalists who contributed to this dissemination, we are particularly thankful to Shayla Love, who wrote the online article entitled: *"A Thousand Years Before Darwin, Islamic Scholars Were Writing About Natural Selection,"* which discussed our papers and their societal repercussions, from influential websites to university courses. Amongst these websites, we are particularly thankful to Nadeem Haque, Savas Konur and Salim Al-Hassani for publicising our work in *Muslimheritage.com*, and to the *American Association of Anatomy* for highlighting the papers in their newsletters, as well as on their website.

Thanks to all these people, websites and organisations, and many others, important stories that have been untold for centuries are now known to millions of people.

We are particularly grateful to Jamil Chishti and Beacon Books for publishing this book, and hope that these stories will now be read, discussed, and disseminated by many millions more.

Preface

"The book is silent as long as you need silence, eloquent whenever you want discourse. He never interrupts you if you are engaged but if you feel lonely, he will be a good companion. He is a friend who never deceives or falters you, and he is a companion who does not grow tired of you."
(Al-Jahiz)

For centuries, Western scholars and historians have tended to promote the ethnocentric idea that many of the major scientific discoveries, or even 'science' itself, originated in the West. In recent decades, this trend of overlooking the contributions of 'others' has, fortunately, changed. Historians of science and scientists are more willing to acknowledge that other societies, including ancient Egyptian, Chinese, Mayan, Arab and Persian, have contributed significantly to scientific discoveries. It is the reason that people from these societies were able to navigate their way across thousands of miles of ocean, predict the movements of stars, and invent things, such as the compass, gunpowder, paper, and means of printing.

However, these achievements have been recognised only in certain disciplines. In other fields of science, such as biology and anatomy, the old ethnocentric views continue to prevail, and it is as if Western scientists and philosophers were the first to have observed the body's soft tissues or discussed ideas about how some species changed with time.

Such narratives were still mainstream some years ago when two Muslim female students came to my lab. Aamina Malik was from a Pakistani family. She was at Howard University studying medicine and wanted to do additional anatomy work by carrying out dissections in my lab. Alongside this research, I asked her to do a detailed literature review on how Muslim scholars had upheld ideas about animal species evolving through time. After all, we already knew that many scholars had, long before Wallace (1823–1913) and Darwin (1809–1882), precisely because the concept of biological evolution—then called 'transmutation'—was already floating around. Aamina was, herself, very reticent about the concept of 'transmutation'. She was of the opinion, acculturated in both Western and most Muslim-majority countries, that many Muslim scholars opposed ideas regarding evolution. According to these narratives, Darwin had 'attacked' God with his heretical ideas. These narratives are not even sophisticated enough to acknowledge that Wallace wrote a manuscript about the theory of evolution by natural selection and sent it to Darwin in 1858, one year before the latter finished his own manuscript and published it in a book entitled *On the Origin of Species*. Thankfully, though, Aamina still agreed to do the work. She had nothing to lose, she told me.

What Aamina's work uncovered was very exciting. It seemed that every week, she was finding more evidence that Muslim scholars and philosophers of the so-called 'Islamic Golden Age'—including many she admired before starting the literature review—had actually defended the idea of some form of transmutation (as we will see in this book). Then, one day, she told me: 'If these Muslim thinkers, who are often considered shining examples of "our" Golden Age, have said these things, they might be true, after all.' Slowly, she became an agnostic in terms of biological evolution, in a way that led her having to defend herself during passionate discussions with her otherwise very supportive family.

The paper that she wrote at this time never had the intention of being the 'last word' or the most detailed account of the transmutation ideas of Muslim scholars before Darwin. Instead, the aim of it—and of this book—is to draw attention to these untold stories. Stories that were often only told by Muslim scholars in centuries gone by, and which have been neglected by others. Stories that have since become taboo, even within Muslim circles, principally due to political interests. By calling attention to such stories, the present book might pave the way for further comparative studies to be undertaken, ones which include more Muslim scholars and philosophers, and which contain detailed comparisons of what scholars and philosophers from other societies were writing at the same time.

To a lesser extent, this also applies to the work done by the other Muslim female student, Malak Alghamdi, who joined our lab back then. Although knowledge about the human body is not as big a taboo within many Muslim communities worldwide as transmutation ideas are, the approach of Western scholars is basically the same. As with the ideas of transmutation by Muslim thinkers before Wallace and Darwin, discoveries about the human body by Muslim scholars before Vesalius (1514–1564), or for that matter after Galen (129–c.210), were almost completely neglected by Westerners, as we shall see.

Although Malak wanted to do a PhD in my lab, conducting work mainly on biological evolution, she told me upfront that she did not want to work on evolution, and could not have the term 'evolution' in any of the publications that came out of her PhD. So, I proposed she should work on human congenital malformations, which would give her experience in anatomy. But again, as a side project, I asked her—being from Saudi Arabia—whether she could do a literature review on the discoveries of Muslim scholars, particularly Arabic and Persian speakers, on the anatomy of the human body, before Vesalius? And she did. Again, her manuscript was not supposed to be the 'last word' on this topic, but rather a wake-up call to historians of science and anatomists, and to pave the way for them to do more research in this area.

And so it did. Firstly, the renowned journal *Anatomical Record*, an official publication of the American Association of Anatomists, not only published the paper on anatomical discoveries, but also selected it for the cover of the issue in which it was published. Then Aamina, Malak and I were invited to give talks about these papers in several universities and countries, including countries with a mainly Muslim population, such as Turkey. This led to a huge dissemination of the papers worldwide, and to the publication of other related news and science articles, of which Shayla Love's *A Thousand Years Before Darwin, Islamic Scholars Were Writing About Natural Selection* was one of the most influential. Websites such as *Muslimheritage.co*, and several university courses included this information in their work. And then, Jamil Chishti from Beacon Books invited us to publish this book, so these untold stories could be accessible to the broader public and thus be read, discussed, and disseminated by more even people. This whole process has had a huge snowball effect since the papers were published in 2017, and this seems to be just the beginning.

However, longstanding biases and prejudices that have been prevalent for many centuries, even millennia, about the 'superiority' of certain societies and cultures over others do not just fade away. A perhaps predictable demonstration that these biases and prejudices continue, indeed prevail, not only in society but also in science and the history of science, is clear from the way some Western scholars have reacted to the papers. They have argued that the transmutation ideas and anatomical discoveries of Muslim scholars before Darwin and Vesalius were 'markedly different' from those of Western scholars in the past few centuries because the Muslim scholars were 'believers'. Such an argument neglects the fact that many Western scholars in the past have been religious, including Newton, Copernicus, and Descartes. Actually, Darwin himself believed in God for a substantial part of his life. Moreover, this claim disregards the fact that the ideas of Western scholars, such as Darwin, were also highly influenced by non-religious *a priori* biases, beliefs and prejudices. In the case of Darwin, he dogmatically believed that women were mentally inferior, and that Europeans were superior and had a 'higher' sense of morality than 'savages'.

However, it should be remembered that 'science' is defined by humans, and all humans have their own biases, prejudices and beliefs; be it in Gods, conspiracy theories, ghosts, the afterlife, or an omniscient Mother Nature. In fact, the evolutionary ideas of Darwin and of many other transmutationists of that epoch, often made reference to an omniscient Mother Nature and referred to natural selection as 'her broom'.

Furthermore, this comparison between the 'high' religious belief of Muslim philosophers and scholars during the Islamic Golden Age versus the religious belief of Western scholars, such as Darwin, can be considered biased in itself, as the former lived centuries before the latter. If we compare how 'fervent' was the religious belief of Darwin and, for that matter, of many Muslim scholars, such as Averroes (**Fig. 2.2**), with that of medieval Western philosophers, such as Tommaso d'Aquino, there is no doubt that, in general, the latter were much more fixated with religion. There are, of course, exceptions to this.

Apart from being based on our two 2017 papers, this book complements them not only with new data and figures, but also with broader discussions on scientific biases and prejudices in Chapter 1. That is: how and why some historians of science and scientists often fell, and still fall, into the trap of bias, racist and ethnocentric narratives, and provides examples which contradict such stories. Chapters 2 and 3 are based on the papers written by our two young, hard-working, and courageous female Muslim students. Lastly, Chapter 4 emphasises how discussions, untold stories and case studies are crucial for a more comprehensive understanding of the history of not only biology, science, and the planet's diverse societies and their philosophies, but also of our way of thinking, biases and prejudices.

We need to start bringing reality, not only to the history of sciences and science itself, but to our societies as a whole.

Rui Diogo, Summer 2021, Washington, DC

Chapter 1

BIASES IN SCIENCE AND HISTORY OF SCIENCES

"Out beyond ideas of wrongdoing and rightdoing there is a field. I'll meet you there."
(Rumi)

Biased Reactions to Untold Stories About Discoveries and Ideas of Muslim Scholars

We have given talks about the topics discussed in this book in several countries. Fascinatingly, the reactions we have encountered, from both the broader public and from scholars, have been remarkably different, illustrating the power of bias in society and also in science. Of course, this should not be news to anybody, particularly not for historians of philosophy and science, yet many still amazingly fall into the trap of such bias. Some Western scholars and historians told us that we should 'know' that the scientific method as defined 'today', originated in Europe some centuries ago. This assertion, however, ignores the role the rest of the world played in developing scientific methods to study the cosmos and the natural world. This typical circular reasoning is used recurrently in ethnocentric biases, beliefs and prejudices.

As will be explained in Chapter 2, most West-

Fig. 1.1. Galen dissecting a monkey, as imagined by Veloso Salgado in 1906.

ern textbooks concerned with the history of anatomy state that Galen (129–c.210)—a Greek physician, surgeon and philosopher during the time of the Roman Empire (**Fig. 1.1**)—was the father of 'human anatomy'. They claim that no major anatomical discoveries about the human body were made after him until the European Renaissance more than a millennium later. Yet, the vast majority of Western historians recognise that the Islamic Golden Age—from the 8th century to the 14th century—was a period of great cultural, economic, and scientific growth in the Muslim world.

Fig. 1.2. Avicenna Portrait on Silver Vase: Museum at BuAli Sina (Avicenna) Mausoleum, Hamadan, Iran.

So, how can it be said that over this time period Muslim scholars did not discover anything new about the human body, especially when we know that scholars, such as Avicenna (980-1037), made major contributions to medicine (**Figs. 1.2, 2.2**). How did they develop medical knowledge without considering the human body, its soft tissues and physiology? This seems rather odd, but biases, prejudices and beliefs are always odd, since they have to be based on, and defend, factually inaccurate ideas.

A typical mechanism used in the construction of biased narratives is to remove agency from 'others' (the outgroup), as if it is only 'us' (the ingroup) who can really be actively and

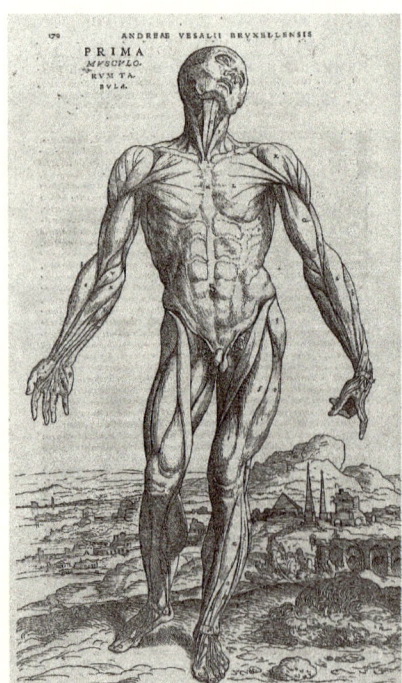

Fig. 1.3. Vesalius' *De humani corporis fabrica* is a superb work with extremely beautiful, and in general very accurate, descriptions of the human body.

purposefully involved in discovery. How could inferior 'others' do that? This mechanism is, in fact, used repeatedly in Western textbooks about the anatomy of history. The existence of the Islamic Golden Age cannot be denied. It is still possible, however, to remove agency from Muslim scholars by saying that their 'great' contribution to anatomy was limited to translating the texts of Greek and Roman scholars, such as Galen, into Arabic or Persian for the purpose of preservation, and these texts only became instrumental during the European Renaissance. In this way, European giants, such as Da Vinci (1452–1519) and Vesalius (1514–1564), could stand on the shoulders of previous ones, such as Aristotle (384–322 BC) and Galen (**Fig. 1.3**). Within this narrative, the giants—those who make new discoveries; the active players—can only be part of the Western ingroup. 'Others', such as Muslim scholars, are merely passive players. They can only translate the Western giants, but never really compete with them or make groundbreaking discoveries of their own.

However, as we will discuss in Chapter 2, historical evidence clearly shows that this narrative is wrong. But what is the reaction when we show such evidence in our talks, through anatomical drawings such as those in **Figs. 2.6–2.8**, and tell the audience that some of the Muslim scholars listed in **Fig. 2.2** knew more about human anatomy than Galen did? Well, that depends on the country and background of the audience.

In Istanbul, Turkey, the vast majority of people in the audience were, as expected, happy and proud. But in European and American cities, we often received harsh, sometimes even aggressive, comments, not only by the lay people, but even local scholars. For instance, following the ethnocentric narrative that science as defined by 'them' was developed in Europe only a few centuries ago, they argued that the anatomical drawings we showed were not truly 'scientific'. When we asked them why, they presented us with vague arguments, including how Muslim scholars did not use the 'scientific method'. This is a rather odd thing to argue because the methodology used to examine human anatomy—in contrast to human physiology, which can involve more experimentation—was basically the same, regardless of whether one was a European in the 1500s or a Muslim scholar in Baghdad centuries earlier. That is, they both learned from observing and/or dissecting dead bodies or from analysing the living who were treated for wounds or undergoing surgery. These observers compiled information in the form of drawings, such as those shown in **Figs. 2.5–2.7**. We know Muslim scholars corrected and updated their records regularly because they often cited older texts, and compared these with the new information they had obtained (**Tables 2.1–2.3**).

In fact, saying that the 'scientific method' was only developed a few centuries ago in Europe minimises what Europeans themselves achieved. For example, how did the Portuguese and Spanish navigate their way to America, not once but numerous times, more than 500 years ago? Was it just luck? Of course not. The expeditions undertaken by Portuguese sailors often involved the use of the mariner's astrolabe, which was an inclinometer used to determine the latitude of the ship during the sea trip by measuring the sun's noon altitude or the meridian altitude of a star with a known declination. The astrolabe was not the product of prayer; it was the product of science.

Well before Europeans, 'others' were navigating their way across the oceans. The ancestors of the inhabitants of modern day Madagascar sailed across the Indian Ocean from the Malay archipelago around the 8th or 9th century. The distance was more than 4,000 miles; further than the trip Christopher Columbus made in 1492 from Castile to the Caribbean. Muslim tradesmen also travelled back and forth between Africa, the Middle East, and Asia, well before Europeans did. The Vikings also went from Europe to North America many centuries before Columbus. Going even further back, the erudite of Ancient Egypt, Ancient China and the Mayan territories, were able to predict—instead of merely observe—the movement of stars, as well as build irrigation systems and pyramids. The Chinese invented the compass, gunpowder, papermaking, and printing. All of these were surely not the product of religious faith, sorcery, or luck. They involved acute observation, repeated trial and error, and the gathering of information. Therefore, the scientific method was not developed a few centuries ago and surely not only by Europeans: this is an unequivocal fact.

Fig. 1.4. One of the many stories about the history of biology that is often untold, even in Western textbooks, is that Wallace—shown here—was the first to finish the writing of a complete manuscript about the theory of evolution by natural selection. He sent that manuscript to Darwin in 1858, that is, one year before Darwin finished his own manuscript about a very similar theory, which he then published in the form of a book: *On the Origin of Species*.

The use of 'special definitions' often exclude evolutionary ideas put forward by 'other' societies. These ideas were similar to those first proposed by Alfred Russel Wallace in his 1858 manuscript and later by Charles Darwin in his 1859 book, *On the Origin of Species* (**Figs. 1.4–1.5**). When we explained in our talks how, centuries before Wallace and Darwin, the writings of some Muslim scholars and philosophers (**Figs. 3.2** and **3.3**) suggested, not only that species change over time, but that even humans could have originated from other species, we were met with criticism from many Western scholars. A typical argument they presented was the use of a 'definition'. That is, the transmutation ideas of such Muslims had nothing to do with Darwinism as 'defined today'. This is, of course, undeniable, since Darwin and Darwinism came many centuries, in some cases more than a millennium, after some of those Muslim scholars and philosophers.

This argument emphasises the typical circular reasoning used in such narratives to exclude 'others', in this case, not only non-European 'others', but even Europeans other than Darwin, such as Wallace. By equating the rise of evolutionary biology with Darwinism, such arguments give the impression that the ideas and observations of other scholars, such as Wallace, were not part of the early history of evolutionary biology; only what Darwin observed, said and wrote was. This type of circular reasoning resembles the way some religious groups or individuals insist that their specific theological interpretations are the sole path to understanding God and the cosmos. It is important to acknowledge that faith traditions naturally hold to their own theological truths while also recognizing the diversity of perspectives on the nature of reality and creation; that is, according to 'our' view and 'our definition'. We will further discuss this topic in Chapter 3.

Another typical argument made by some Western scholars is that the transmutation ideas of the Muslim scholars and philosophers were 'markedly different' from those defended by Western scholars in the last few centuries because those Muslims were 'believers'. As noted in the Preface, this argument is based on a comparison between how 'high' the

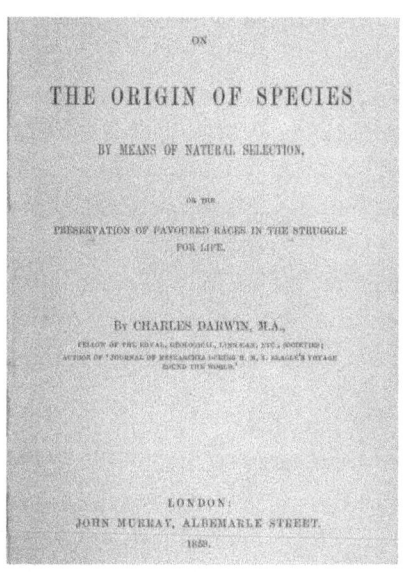

Fig. 1.5. Charles Darwin's 1859 *On the Origin of Species* compiles many fascinating facts about the natural world to support the theory of evolution by natural selection.

religious belief of Muslim philosophers and scholars was during the Islamic Golden Age versus the religious belief of Western scholars, such as Darwin. However, this cannot be thought of as a direct comparison as the former lived centuries before the latter. If we would compare how 'fervent' was the religious belief of Darwin, and for that matter, of many Muslim scholars, such as Averroes, with that of medieval Western philosophers, there is no doubt that, in general, the latter were much more obsessed with religion, although there are of course exceptions. To put this into perspective, Averroes, a Muslim polymath who wrote about subjects as diverse as philosophy, theology, medicine, astronomy, physics, psychology, mathematics, linguistics, Islamic jurisprudence and law, and anatomy, was born in Córdoba, Spain, in 1126. This was almost a century before Tommaso d'Aquino, a Western philosopher, theologian and jurist, who was born in Sicily, in 1225.

Moreover, such arguments ignore the fact that even several centuries later most Western scholars were religious, including Newton, Copernicus, and Descartes. Actually, Darwin himself believed in God for a substantial part of his life. Furthermore, this argument also neglects the fact that the ideas of Western scholars, such as Darwin, were highly influenced by other, non-religious, *a priori* biases, beliefs, and prejudices. For instance, in the case of Darwin, he dogmatically believed that women are mentally inferior to men, and that Europeans are superior and have a 'higher' morality than so-called 'savages'. These beliefs led him to construct inaccurate 'evolutionary facts' in his 1871 book, *The Descent of Man*, to 'scientifically' support such biases. As noted in the Preface, science is defined by humans rather than super-humans or supernatural deities, so obviously all scholars—including ourselves, the authors of this book—have their own biases, prejudices and beliefs; be they in Gods, conspiracy theories, ghosts, the afterlife, or an omniscient Mother Nature. In fact, the evolutionary ideas of Darwin and of many other transmutationists of that epoch, often refer to such an omniscient Mother Nature, and to natural selection as Her 'broom'.

Darwin's Own Beliefs, Biases and Prejudices

I have discussed extensively Darwin's own beliefs, biases and prejudices in a recent book, entitled *Racism, Sexism, and Darwin's Idolization – and their tragic scientific and societal repercussions until today*. Therefore, here I will provide just a brief discussion of those issues that relate directly with the topics analysed in the present book. An autobiographical chapter written by

Charles Darwin for his family, together with a collection of his letters, was published five years after he died, in 1887, by his son Francis Darwin, under the title, *The Life and Letters of Charles Darwin*. In that autobiographical chapter, Darwin wrote:

> After having spent two sessions in Edinburgh, my father perceived, or he heard from my sisters, that I did not like the thought of being a physician, so he proposed that I should become a clergyman... Considering how fiercely I have been attacked by the orthodox, it seems ludicrous that I once intended to be a clergyman. As it was decided that I should be a clergyman, it was necessary that I should go to one of the English universities and take a degree; but as I had never opened a classical book since leaving school, I found to my dismay, that in the two intervening years I had actually forgotten, incredible as it may appear, almost everything which I had learnt, even to some few of the Greek letters. I did not therefore proceed to Cambridge at the usual time in October, but worked with a private tutor in Shrewsbury, and went to Cambridge after the Christmas vacation, early in 1828.
>
> During the three years which I spent at Cambridge my time was wasted, as far as the academical studies were concerned, as completely as at Edinburgh and at school. I attempted mathematics, and even went during the summer of 1828 with a private tutor (a very dull man) to Barmouth, but I got on very slowly. The work was repugnant to me, chiefly from my not being able to see any meaning in the early steps in algebra. This impatience was very foolish, and in after years I have deeply regretted that I did not proceed far enough at least to understand something of the great leading principles of mathematics, for men thus endowed seem to have an extra sense. But I do not believe that I should ever have succeeded beyond a very low grade. With respect to Classics I did nothing except attend a few compulsory college lectures, and the attendance was almost nominal.
>
> In my second year I had to work for a month or two to pass the Little-Go, which I did easily... In order to pass the B.A. examination, it was also necessary to get up Paley's *Evidences of Christianity* and his *Moral Philosophy*. This was done in a thorough manner, and I am convinced that I could have written out the whole of the *Evidences* with perfect correctness, but not of course in the clear language of

Paley. The logic of this book and, as I may add, of his *Natural Theology*, gave me as much delight as did Euclid. The careful study of these works, without attempting to learn any part by rote, was the only part of the academical course which, as I then felt and as I still believe, was of the least use to me in the education of my mind. I did not at that time trouble myself about Paley's premises; and taking these on trust, I was charmed and convinced by the long line of argumentation.

By answering well the examination questions in Paley, by doing Euclid well, and by not failing miserably in Classics, I gained a good place among the oi polloi or crowd of men who do not go in for honours... But no pursuit at Cambridge was followed with nearly so much eagerness or gave me so much pleasure as collecting beetles. It was the mere passion for collecting, for I did not dissect them, and rarely compared their external characters with published descriptions, but got them named anyhow... Public lectures on several branches were given in the University, attendance being quite voluntary; but I was so sickened with lectures at Edinburgh that I did not even attend Sedgwick's eloquent and interesting lectures. Had I done so I should probably have become a geologist earlier than I did.

One very interesting point about Darwin's autobiography is that Darwin himself deconstructs the notion that he was a polymath 'sage'—an image that was constructed by subsequent scholars and which endures in popular culture to this day. In his autobiography he explains that, for him, many fields of knowledge were 'dull'. In his account of the time he spent at Cambridge, he was far more interested in collecting beetles. Such passion in the natural world is repeated countless times in books idolising Darwin. Undoubtedly, he was a skilled biologist and naturalist; indeed, one of the top naturalists of all time. His detailed observations of the natural world, in particular of non-human organisms, were truly outstanding, and factually accurate. However, his lack of interest in other fields of knowledge, before his travels on the Beagle, meant that he did not reach the same level of detailed, and accurate, observations of humans themselves, and of human evolution. One could say that, as an anthropologist, Darwin was much less skilled, being less objective and much more biased. This is clearly evidenced in the parts about human evolution in his books, *Descent* (1871) and *Expression* (1872), in which he stated that 'savages' were morally inferior to Victorians. He argued that the

Victorians' distinct societal hierarchies and gender roles were the pinnacle of social evolution, and that women were 'naturally' mentally inferior to men. Most of the 'observations' he made and 'evolutionary facts' he constructed about the societies he encountered whilst travelling on the Beagle were not a reflection of what he truly saw, but of the *a priori* negative *image* of 'savages' he had been acculturated to believe.

The lack of depth and critical reasoning on such complex societal matters, as well as the acculturation process itself (including the propagation of systemic racism and sexism), enabled him to view his Victorian society as 'good'—especially in the moral sense. In comparison, the 'others' were viewed as inferior and less moral, or more 'brutish'. It is, perhaps, ironic of Darwin to 'conclude' that Victorian society was morally 'better' and more 'evolved' than 'other' human societies, as this assertion contradicts his claim in *On the Origin of Species*, that, generally, there is no 'better' or 'worse' in evolution, but just local adaptations to local environments. Such biased 'evolutionary conclusions' were not the result of scientific observations and the use of the scientific method, but mainly the outcome of Darwin's own biases, prejudices and beliefs. In other words, he, as with many humans and scholars,

Fig. 1.6. The Yaghan people are one of the indigenous groups of the Southern Cone, who are regarded as the southernmost peoples in the world. In the 19th century they were known as Fuegians by the English-speaking world because their traditional territory includes the islands south of Isla Grande de Tierra del Fuego, extending their presence into Cape Horn.

Fig. 1.7. Jemmy Button in 1831 and 1834.

embraced the biased beliefs and narratives that he heard at home, school, and from people around him as a given, without actively questioning them; not even when they were contradicted by many of the things he saw before his eyes in his Beagle voyage. Instead, these biases were camouflaged as 'scientific observations' and 'evolutionary facts' about the 'others'—i.e., the 'savages'—that he encountered on that voyage, such as the Fuegians from Tierra del Fuego in the southern region of South America (**Figs. 1.6 and 1.7**):

> The Captain sent a boat with a large party of officers to communicate with the Fuegians... When we landed the party looked rather alarmed, but continued talking and making gestures with great rapidity. It was without exception the most curious & interesting spectacle I ever beheld. I would not have believed how entire the difference between savage and civilized man is. It is greater than between a wild and domesticated animal, inasmuch as in man there is greater power of improvement. The chief spokesman was old and appeared to be head of the family; the three others were young powerful men and about 6 feet high. From their dress they resembled the representations of *Devils on the Stage*, for instance in *Der Freischutz*. The old man had a white feather cap; from under which, black long hair hung round his face. The skin is dirty copper colour... the only garment was a large guanaco skin, with the hair outside. This was merely thrown over their shoulders, one arm and leg being bare; for any exercise they must be absolutely naked... Their very attitudes were abject, and the expression distrustful, surprised and startled: Having given them some red cloth, which they immediately placed round their necks, we became good friends.
>
> This was shown by the old man patting our breasts and making something like the same noise which people do when feeding chickens. I walked with the old man and this demonstration was repeated between us several times: at last he gave me three hard slaps on the

breast and back at the same time, and making most curious noises. He then bared his bosom for me to return the compliment, which being done, he seemed highly pleased. Their language does not deserve to be called articulate: Capt. Cook says it is like a man clearing his throat; to which may be added another very hoarse man trying to shout and a third encouraging a horse with that peculiar noise which is made in one side of the mouth. Imagine these sounds and a few gutterals mingled with them, and there will be as near an approximation to their language as any European may expect to obtain... They are excellent mimics; if you cough or yawn or make any odd motion they immediately imitate you. Some of the officers began to squint and make monkey like faces; but one of the young men, whose face was painted black with white band over his eyes was most successful in making still more hideous grimaces. When a song was struck up, I thought they would have fallen down with astonishment; and with equal delight they viewed our dancing and immediately began themselves to waltz with one of the officers... If their dress and appearance is miserable, their manner of living is still more so... I believe if the world was searched, no lower grade of man could be found.

Were Darwin's ideas about the Fuegians, and non-European people in general, a *tabula rasa* before he actually encountered them? Were the 'facts' he wrote about them based solely on his 'objective' observations of them? Of course not. Before Darwin's travels, Malthus—who profoundly influenced Darwin—looking for an example of the world's most downtrodden 'savages', had already written about 'the wretched inhabitants of Tierra del Fuego'. These peoples had been said by earlier European travellers to be 'at the bottom of the scale of human beings'. When Darwin arrived in Tierra del Fuego, he agreed with Malthus, thereby confirming his own *a priori* biases, as seen in the above excerpt. The way that Darwin *perceived* the 'savages', and the 'scientific facts' he wrote about them, describing them as 'brutish' and 'barbaric' and 'cannibals', had nothing to do with reality, but were, instead, steeped in his own *beliefs*. As noted by Desmond and Moore, in their 1991 book, *Darwin*:

> The scenery was spectacular. They proceeded along the Beagle Channel, tracking a granite ridge of mountains, the backbone of Tierra del Fuego... While dining on shore half a mile away, they

heard a 'thundering crash' as a huge mass of ice fell from its face. The impact sent 'great rolling waves' racing towards their flotilla. Darwin was quick to act. He and others seized the boats, hauling them to safety just as the first breaker crashed down... But Darwin had acted less out of bravado than fear. Without the boats, he reflected, 'how dangerous would our lot have been, surrounded... by **hostile Savages** and deprived of... provisions.' Some Fuegians were indeed hostile. With courage 'like that of a wild beast', they menaced the party's overnight camp, and armed guards were posted. Charles, keeping watch, shivered at his vulnerability in this land. 'The quiet of the night is only interrupted by the heavy breathing of the men and the cry of the night birds—the occasional distant bark of a dog reminds one that the Fuegians may be prowling, close to the tents, ready for a fatal rush.'

Darwin and his fellow travellers *believed* they were among cannibals. However, texts which were written by Christians who lived amongst the Fuegians, as well as observations by archaeologists and anthropologists who have explored the region, strongly indicate that they were *not* living among cannibals. Thus, it is safe to assume that Darwin never actually *saw* any act of cannibalism within the Fuegian peoples, but opted to *believe*, instead, in the horror stories told, not only by Europeans, but also Fuegians—such as Jemmy Button (**Fig. 1.7**)—brought over by English explorers to be 'civilized' in London.

What one hears about 'others', especially if they are derogatory and confirm *our* ideas about *our* superiority, can be more important than what one actually observes. Such conduct is carried out by all societies and groups towards 'others', in a manner known as *tribalism*. However, no society or culture is biologically 'good' or 'bad' or 'better' or 'worse': these terms are just social constructions that have nothing to do with the reality of the natural world.

In a nutshell, instead of recording facts based on the observation of 'others' and elaborating a scientific theory about human evolution based on them, Darwin constructed inaccurate 'facts' about certain human societies to support the Victorian biases, fears, and tales that he, like many other—but not all—Victorians blindly accepted *a priori*. To paraphrase George Orwell, *to see what is in front of one's nose needs a constant struggle*. It could be argued that it is *because of*—not *despite*—the ethnocentric and racist 'evolutionary facts' Darwin wrote about humanity and human evolution, that his books became, and continue to be, widely cited and revered in literature, particularly by those who are portrayed as the evolutionary 'winners' in them, i.e. Westerners and Western scholars. These are the same scholars who have systematically minimised and neglected many of the important concepts on transmutation and scientific discoveries, such as anatomical findings, made by Muslim scholars and philosophers.

It is perhaps not a coincidence that, apart from *On the Origin of Species*, it is Darwin's earlier two books that are today celebrated by Western scholars, as evidenced by the commemorations marking the 150th anniversary of the first edition of *The Descent of Man*. We did not see such commemorations on the 150th anniversary of his monograph, *The Structure and Distribution of Coral Reefs*, published in 1842, nor for his monograph, *Volcanic Islands*, which was published in 1844. The marked difference in the way Darwin's biased and scientifically inaccurate books are idolised in comparison to his brilliant and scientifically accurate works on barnacles, volcanoes, coral reefs, and worms, tells us a great deal about the power of human, societal and scientific preconceptions. This is a logical consequence within the type of dogmatic and circular self-reinforcing type of reasoning that is involved in the construction, propagation, and acculturation of biased narratives and beliefs.

This is an important point, because the issue is not Darwin's problematic ideas, but principally the type of people who idealise him and use his ideas to justify their own controversial views of humanity. Darwin's observations have provided hugely powerful weapons to politicians and such, who have used them to bolster their own views on 'others' and women. If Darwin himself, as well as his ideas—including his most flawed racist and sexist ones— had not been idealised, idolised, and worshipped by so many scientists, historians, and scholars, political leaders and other social actors would not have been able to 'scientifically' justify their own incorrect notions and policies. In this respect, the role played by historians and scientists can be strongly criticised. Scholars such as these should know better than others how biases and contingencies of life often influence scientific ideas, and, accordingly, try to avoid falling into the trap that Darwin himself did.

In this sense, it is important to note that, while *On the Origin of Species* is less biased and more scientifically accurate than *The Descent of Man* and *The Expression of the Emotions in Man and Animals*, it still includes 'evolutionary facts' that stem from Darwin's own biases and beliefs. This is because, despite not referring to humans directly, its content was highly influenced by anthropocentric concepts. This can be seen in the exaggerated parallels

Darwin draws between artificial selection—purportedly developed by humans—and natural selection, as well as the repeated extrapolation of biological evolution from a Victorian narrative, for instance, capitalistic ideas about competition and selfishness within a brutal 'struggle-for-existence'. This bias was recognised by Karl Marx, who noted that: 'it is remarkable how Darwin rediscovers among beasts and plants the society of England, with its division of labour, competition, opening up of new markets, inventions, and the Malthusian struggle for existence.'

This 'struggle for existence' refers to a concept developed by Thomas Robert Malthus, which highly influenced Victorian society and Darwin's works: namely that a population will increase exponentially if unchecked, while the resources tend to only increase arithmetically, so the checks that exist on population growth result in a brutal struggle for existence. Or, as Nordenskiold put it in his *The History of Biology: A Survey*, 'From the beginning, Darwin's theory was an obvious ally to liberalism; it was at once a means of elevating the doctrine of free competition, which had been one of the most vital cornerstones of the movement of progress, to the rank of a natural law, and similarly the leading principle of liberalism, progress, was confirmed by

Fig. 1.8. Painting *Death of Hypatia in Alexandria*. Hypatia was a brilliant public speaker and scholar and wrote on mathematics and astronomy, inventing the astrolabe for ship navigation and devices for measuring the density of fluids.

the new theory.' That is, 'the deeper down the origin of human culture was placed, the higher were the hopes that could be entertained for its future possibilities. It was no wonder, then, that the liberal-minded were enthusiastic; Darwinism must be true, nothing else was possible.'

However, the crude reality is that this is *not* right. One example of this concerns the evolutionary 'facts' about women which Darwin writes about in *The Descent of Man*. In it, and other writings, he states that women have 'naturally' lower intellectual powers, imagination and ingenuity than men, and that women 'have never advanced the science'; the implication being that women's inferior mental skills and 'ignorance', as well as their 'nature' in general, did not help them to 'naturally' become top scientists.

All these 'facts' are of course biologically *and* historically wrong. Women are not mentally inferior to men. Even at the time he wrote such sexist statements, many women had become prominent scientists, writers with powerful imaginations, and successful leaders—including Queen Victoria, who ruled his country for most of his life. One example of a prominent scientist is Hypatia, the renowned Hellenistic philosopher, mathematician and astronomer (**Fig. 1.8**). She was born in 360 AD and lived in Egypt, then part of the Eastern Roman Empire, where she taught astronomy and philosophy. Darwin did not refer to her or any other female scientists, nor did he even consider the numerous reasons that prevented countless, similarly bright, women from becoming leading scientists, writers, and so on. For instance, he did not mention that, during much of his lifetime (1809–1882), women could not go to university in England. Only in 1868 did the University of London's Senate vote to admit women to sit the General Examination, thereby becoming the first English university to accept women. While Darwin had the privilege of going to university before 1868—in fact to two prestigious universities: Edinburgh and Cambridge—*all English women* were forbidden from doing so, even if they were extremely motivated and capable.

Darwin's Mother Nature, Her 'Broom', and the Notion of 'Progress'

Another powerful example of how Darwin's ideas about biological evolution were influenced by what he read and, importantly, did *not* read, before his travels on the Beagle concerns his teleological views on Mother Nature. Teleological narratives are related to the notion that things have a 'special' or 'cosmic' purpose or use. The word 'teleology' builds on the Greek *telos*, meaning purpose or end, and *logos*, meaning explanation or reasoning. As we see from his autobiography, Darwin recognised that: 'in order to pass the B.A. examination, it was also necessary to get up William Paley's *Evidences of Christianity* and his *Moral Philosophy*. This was done in a thorough manner, and I am convinced that I could have written out the whole of the *Evidences of Christianity* with perfect correctness.' As he wrote: 'the logic of this book and, as I may add, of his *Natural Theology*, gave me as much delight as did Euclid. The careful study of these works, without attempting to learn any part by rote, was the only part of the academical course which, as I then felt and as I still believe, was of the least use to me in the education of my mind. I was charmed and convinced by the long line of argumentation.'

This fascination that Darwin had at Cambridge with William Paley is key to understanding his broader view of the natural world and its evolution. This is because Paley, and in particular, his 'watchmaker analogy', is one of the most famous examples of teleological thinking. The 'watchmaker' analogy argues that a design implies a designer. If you were to find a watch on the ground you would assume that it was made by an intelligent designer—that is, by a human. Similarly, the only way to explain the origins of humanity is to assume that it was made by a creator deity. Humans, even those that declare themselves to be 'non-believers', tend to think about final purposes: things happen for a reason. With regards to science and philosophy, 'nature does nothing in vain' was a notion defended by Aristotle thousands of years ago. We hear this in documentaries about plants, animals, fungi, bacteria, and non-living organisms such as viruses; as attested by the recent Covid-19 pandemic. Such narratives are plagued by teleological terms, just as they were a millennia ago. Even in popular documentaries about the natural world, such as those narrated by David Attenborough, we hear how 'the wings of birds were made to fly'. This type of circular reasoning resembles the way some religious groups or individuals insist that their specific theological interpretations are the sole path to understanding God and the cosmos.

Secondly, the original function of tissues that went on to form wings, which were conferred to the ancestors of modern birds, was not that of flying. The wings of birds include bones and muscles which were already present in their non-winged ancestors. Moreover, the very first structures that could be called 'wings' were probably linked to another function, such as thermoregulation; that is, the ability to keep the body temperature within certain boundaries, even when the surrounding temperature is different. This is what biologists call an 'exaptation': a biological structure that originally relates to function A adapts so that it becomes useful as function B. Such functional shifts are common in evolution and illustrate the complexity of how anatomical structures develop over time. While some interpretations may see purpose in these changes, the scientific perspective emphasizes gradual adaptation rather than purposeful design.

However, contrary to what is often claimed about Darwin, he did not completely remove teleology from the natural world by demonstrating that a creator was not necessary to explain the origins of organisms, including humans. In Darwin's time, Paley's analogy played a crucial role in the development of natural theology—a field that seeks to understand the divine through the study of nature. Natural theology offers arguments for God's existence based on reason and observation of the natural world, reflecting the belief that natural phenomena demonstrate intentional design. Today, this line of reasoning is also used to support arguments for intelligent or purposeful design in the cosmos. Proponents of this perspective often cite the apparent 'fine-tuning' of natural processes as evidence of a creator's guiding hand.

A huge paradox, or logical inconsistency, is that, while Darwin removed the religious teleological 'watchmaker' analogy from the scientific field of biology, he was unable to remove teleology from the way he—and thus his followers—saw the natural world. For instance, as noted above, regarding human evolution and human 'races', Darwin often used terms that were in vogue in Victorian society, such as the 'progress' from 'lower' to 'higher' groups of societies or morality. Such terms were, in turn, directly related to the much older teleological notion of *scala naturae* that Ancient Greek thinkers, such as Aristotle, had put forward millennia before

Fig. 1.9. *Ascent of Life* by F. Besnier, 1886, illustrating the notion of a ladder of life.

(**Fig. 1.9**). This concept of *scala naturae,* also known as the 'chain of being' or 'ladder of life', was, and still is, extremely important in Western thought. It influenced Muslim scholars and philosophers as well, who read and quoted Greek and Roman writers extensively, as discussed in the next chapters. In fact, this concept of the chain of being is crucial in understanding the idea of transmutation amongst those Muslim writers in **Figs. 3.2** and **3.3**.

Lovejoy's book *The Great Chain of Being* (first published in 1936), provided a superb introduction to this concept, its related notion of 'progress', and of teleology in general, with regards to Western philosophy and science. As he explained, in the chain of being there is a 'progress' from lower forms, such as plants, to non-human animals and then to humans, who are the culmination point of such a 'progression' towards perfection. This notion is profoundly connected with our quest towards understanding humankind, including its 'subgroups', or imaginary 'races', and their place in nature.

In more recent years, Scott Turner's 2007 publication, *The Tinkerer's Accomplice,* provides a succinct historical background to the idea of intentionality in nature, referring, for example, to platonic teleology, in which crabs were made by God to offer particular moral examples as a means of teaching humans. Aristotle, who greatly influenced Darwin, saw purpose in a different, more dynamic way by focusing on physiology. For instance, Aristotle suggests that every animal has a purpose, and when removed from that context, such as a fish taken out of water, it will intentionally try to return to it or will adapt to its new niche, for example, by adjusting to living on land. However, it should be noted that although Ancient Greek scholars, such as Aristotle, accepted this notion of the chain of being, it does not mean that the chain (shown in **Fig. 1.9**) had literally occurred during the existence of life in this planet through transmutation, that is, via biological evolution. Most Greek, Roman, and Muslim scholars and philosophers, as well as some pre-Wallace and pre-Darwin ones, did not refer to transmutation as a process that really happened during the history of life on Earth. Rather, they considered it a theoretical idea of what was in the mind of the Gods when they created the natural world; for instance, humans being the 'higher' organisms. But some scholars and philosophers—including some Muslim ones—asserted that transmutation actually did occur during the history of life on Earth, either 'naturally' i.e., without direct intervention of the Gods, or as part of their masterplan.

The case of Darwin is interesting because, on the one hand, he defended the idea that transmutation truly happened and continues to happen, and stated that this had nothing to do with the masterplan of any God. On the other hand, he repeatedly used teleological terms, such as 'progress', 'higher', 'superior' or 'better', particularly about human evolution, but sometimes also about biological evolution, as noted above. Moreover, he often referred to Mother Nature as an omniscient individual being or deity. In *On the Origin of Species*, Darwin wrote that Mother Nature 'is daily and hourly scrutinizing, throughout the world, every variation, even the slightest; rejecting that which

Fig. 1.10. Erasmus Darwin, Charles' grandfather, also defended evolutionary ideas, some of them very similar to those later defended by his grandson.

is bad, preserving and adding up all that is good.' Apart from using terms such as 'higher and lower forms of life', Darwin wrote in his *Notebook B* that 'there must be progressive development; for instance, none of the vertebrata [animals with vertebrae] could exist without plants and insects had been created; but on the other hand, creations of small animals must have gone on since from parasitical nature of insects and worms.' At first, it could seem paradoxical that Charles Darwin was so fascinated and influenced by Paley's teleological 'argument from design' and repeatedly used teleological terms in his evolutionary works. This is because some of Paley's writings were a response to the evolutionary ideas of Charles Darwin's own grandfather, Erasmus Darwin (**Fig. 1.10**), particularly those found in his two-volume work, *Zoonomia*. However, if we dig a little deeper, it becomes apparent that this is not a paradox at all. As noted in Howard E. Gruber's 1981 book, *Darwin on Man*, Erasmus was also fascinated by the 'argument from design'. The key difference between Erasmus' writings and those of Paley, was that the former mainly ascribed such 'design' to Mother Nature, precisely as his grandson, Charles, did decades later:

> Both Darwins, but especially the grandson, were deeply influenced in their view of change and struggle by the argument from design. [Within] this teleological view... the entire course of evolution was seen as a series of small readjustments on the part of a self-regulating system, nature as a whole... The Darwins were not Natural Theologians, but they did try to study nature as a whole.

And they both had an attitude toward nature that might be called worshipful—reverential, enthusiastic, and poetic... In a very general sense, both Erasmus and Charles accepted the utilitarian ethic... For Erasmus... 'the greatest good for the greatest number' was translated into a 'greatest happiness principle', about which he wrote on more than one occasion.

In *Zoonomia* he expressed his belief 'in the progressive increase of the wisdom and happiness' of the inhabitants of the earth... In the *Temple of Nature*, he wrote that 'the sum total of the happiness of organized nature is probably increased... when one large old animal dies, and is converted into many young ones...' [His grandson] Charles discussed the sources of happiness in the M notebook. His argument there is similar to his grandfather's. Higher thought processes yield greater happiness... perhaps too hopefully, he writes of aggressive impulses that 'with lesser intellect they might be necessary and no doubt were preservative and are now, like all other structures slowly vanishing.' He sounded the same note of evolutionary optimism in *Descent*: in discussing the conflict in each between social and anti-social tendencies, he concluded that through natural evolutionary processes 'the struggle between our higher and lower impulses will be less severe, and virtue will be triumphant.'

In other words, Charles wrote notes to remind himself that he should not fall into the trap of teleology, but it seems this was inevitable when he writes of 'biological progress', 'higher and lower impulses', and that 'virtue will be triumphant'. So, we may ask: *whose virtue is he talking about*? Of humans, as the 'higher' animal? Or bacteria? Or perhaps it is viruses, such as Covid-19? Additionally, we can question why it is that virtue should be ultimately triumphant in nature? Is it possible for there to be a direction towards 'virtue', if his theory of natural selection relates to local adaptations that depend on specific physiological, ecological, geographic, and environmental conditions, which, in turn, are partially related to chance and contingency? What a coincidence it would be if within all those different environmental conditions, in every continent, and amongst all the species that there ever were, 'virtue' would always 'be triumphant' and organisms would be 'happier'.

What we see in Darwin's writings is, of course, not unique, but instead they merely a reflect a theme that is recurrent in science, and which is a central topic of the present book: namely the profound link between teleology, biases, prejudices, and scientific ideas. Later in life Darwin himself admitted that he was very 'pleased' to be recognised as a teleological writer. In 1874, he wrote a letter to Asa Gray—who had published an article in the scientific journal *Nature* that same year paying tribute to Darwin—stating: 'What you say about Teleology ["let us recognise Darwin's great service to Natural Science in bringing back to it Teleology: so that, instead of Morphology versus Teleology, we shall have Morphology wedded to Teleology"] pleases me especially, and I do not think any one else has ever noticed the point... I have always said you were the man to hit the nail on the head.'

In short, contrary to those scholars who argue that Darwin removed teleology from nature, it appears that Darwin himself recognised that he had 'wedded' biological fields, such as anatomy and evolutionary biology, to teleology. Nowadays, this tradition is followed by not only many anatomists, but also by a large number of evolutionary biologists and evolutionary psychologists. Such scholars continue to blindly accept a teleologically adaptationist way of thinking. The term 'adaptationism' is a Darwinian one that views many anatomical or psychological traits of organisms as being evolutionary adaptations. In the 20th century, many evolutionaries began to defend a more extreme adaptationist view of evolution than that defended by Darwin himself, arguing that the vast majority of traits are adaptations. Although this has changed in recent decades, some scholars, particularly in the fields of evolutionary medicine, evolutionary psychology, and behavioral ecology, still hold on to this position. The problem with this is that such extreme versions of adaptationism often involve a non-falsifiable (and, therefore, a quasi-religious) type of circular reasoning in which traits are seen as 'useful adaptations' *a priori*. According to such a preconceived notion, the usefulness of a certain feature just needs to be proven with time and, consequently, cannot truly be disproved; as astutely pointed out by Stephen Jay Gould.

Interestingly, Darwin recognised that he often fell into the teleological trap of personifying Mother Nature. In the variorum edition of *On the Origin of Species* (he produced six editions of that book during his lifetime), he wrote: 'so again it is difficult to avoid personifying the word Nature; but I mean by Nature, only the aggregate action and product of many natural laws, and by laws the sequence of events as ascertained by us... with a little familiarity such superficial objections will be forgotten.'

The problem is that the personalised and teleological way in which Darwin and scholars, including evolutionary biologists, refer to Mother Nature continues to prevail today, particularly within media and popular culture.

For those readers that want to know more about this topic, one of the authors of this work (Diogo) has recently discussed these issues in a book entitled, *Darwin's Racism, Misogyny and Idolization - and Their Tragic Societal and Scientific Repercussions Until Today.*

Other Examples of Political and Societal Influences on Science and the History of Science

In the last section of this chapter, I will provide examples (selected from numerous case studies) of the critical role played by societal and political biases within science and the history of science. Homosexuality is one such example; having been the subject of many scientific biases and prejudices. For a long time, the evolution of homosexuality was not widely researched within scientific publications, and when it was, the focus was exclusively on human homosexuality. This discussion has only become broader in recent times with publications such as *Homosexual Behavior in Animals – an Evolutionary Perspective*, edited by Sommer and Vasey and published in 2006:

> Earlier studies of animal behaviour tended to dismiss occurrences of same-sex sexual behaviour as mere quirks or such instances were classified as pathological manifestations. The use of caged subjects was prevalent and meant that these interactions were invariably characterized as abnormal products of captivity, unlikely to be found in 'nature'. As early as the 1700s biologists such as George Edwards (1758–64) were speculating on the causes of such behavioural 'abnormalities'. He stated that 'three or four young [bantam] cocks remaining where they could have no communication with hens... each endeavoured to tread his fellow, though none of them seemed willing to be trodden. Reflections on this odd circumstance hinted to me, why the natural appetites, in some of our own species, are diverted into wrong channels'... Nevertheless, more and more detailed studies of animals in their natural environments made it increasingly difficult to discount all sexual interactions in animals among members of the same sex as exceptions, as idiosyncrasies, or as pathologies... A recent encyclopedic volume by Bruce Bagemihl (in 1999) on animal homosexual behaviour provides evidence that hundreds of mammals, birds, reptiles, amphibians, fishes, insects, spiders and other invertebrates engage in same-sex sexual activity. Clearly, what

was once thought to be an aberration appears to be a behavioural pattern that is broadly, albeit unevenly, distributed across the animal kingdom... Indeed, within a select number of species, homosexual activity is widespread and occurs at levels that approach or sometimes even surpass heterosexual activity.

From a scientific perspective, homosexual behavior is observed in many non-human animal species, indicating that it is not uniquely human. While religious interpretations may vary, scientific studies have documented such behaviors as part of natural animal diversity. However, a major problem that continues to plague such research on homosexuality concerns another teleological narrative, namely the quest by many scientists to seek a 'purpose'. This particularly applies to adaptationists who want to find, at all costs, an adaptive advantage—very often a current one—for every trait found in living beings. However, we now know that many traits can be present within any taxa without providing an advantage. Unfortunately, such ideas continue to plague evolutionary research, particularly fields such as evolutionary psychology. Although the authors of the above-quoted book recognise the danger of falling into the 'adaptationist trap', they, nevertheless, do so when attempting to explain the 'purpose' of homosexuality in the animals they study. Numerous 'purposes' are proposed in the book based on the *a priori* assumption that homosexuality might be an evolutionary adaptation—a clear illustration of circular adaptationist reasoning—despite the obvious fact that same-sex intercourse does not lead to the production of offspring. Such 'purposes' include 'controlling population size', 'dominance expression', 'social tension regulation', 'reconciliation', 'social bonding', 'alliance formation', 'acquisition of alloparental care', 'mate attraction', 'inhibition of competitor's reproduction', 'kin selection', and 'practice for heterosexual activities'.

Another 'other' that is often attacked in biased scientific narratives, including those published by Darwin, is 'women'. Mostly constructed by men, these narratives 'naturalize' women's supposed 'inferiority' or 'lack of mental abilities', and their 'tendencies to have domestic roles'. For instance, in *Marriage – a History: How Love Conquered Marriage*, Stephanie Coontz explains that:

According to the protective or provider theory of marriage... still the most widespread myth about the origin of marriage... women and infants in early human societies could not survive without the men to bring them the meat of woolly mammoths and protect them from marauding saber-toothed tigers and from other men seeking to abduct them. But males were willing to protect and provide only for their 'own' females and offspring they had a good reason to believe were theirs, so a woman needed to find and hold on to a strong, aggressive male.

One way a woman could hold a mate was to offer him exclusive and frequent sex in return for food and protection. According to the theory, that is why women lost the estrus cycle that is common to other mammals, in which females come into heat only at periodic intervals. Human females became sexually available year-round, so they were able to draw men into long-term relationships...

[However,] studies of actual human hunting and gathering societies also threw doubt on the male provider theory—in such societies, women's foraging, not men's hunting, usually contributes the bulk of the group's food. Nor are women in foraging societies tied down by child rearing. One anthropologist, working with an African hunter-gatherer society during the 1960s, calculated that an adult woman typically walked about twelve miles a day gathering food, and brought home anywhere from 15 to 33 pounds. A woman with a child under two covered the same amount of ground and brought back the same amount of food while she carried her child in a sling, allowing the child to nurse as the woman did her foraging. In many societies women also participate in hunting, whether as members of communal hunting parties, as individual hunters, or even in all-female hunting groups. Today most paleontologists reject the notion that early human societies were organized around dominant male hunters providing for their nuclear families. Instead there is strong evidence that in many societies, [particularly] sedentary agriculturalists, marriage was indeed a way that men put women's labor to their private use. Women's bodies came to be regarded as the properties of their fathers and husbands.

Within a patriarchal society such narratives about marriage and 'women's natural tendency to do domestic tasks' continues to have dramatic implications. In the 21st century, even in so-called 'developed' countries, such as the Netherlands, U.S.A., and Germany, women, on average, have less leisure time than men, as shown in **Fig. 1.11**; the one exception to this being Norway. In essence, what Coontz argues is that this 'man-the-provider' theory is an outdated, pseudo construct based on misogynistic biases. This argument is backed up by studies and literature reviews presented in various books and papers; books, such as Sarah Blaffer Hrdy's influential 2009 book, *Mothers and Others,* which highlights these 'man-made' tales, including some of Darwin's ideas:

> From the outset, they [evolutionists] assumed... that [the] provider must have been her [the wife's] mate, as Darwin himself opined in *The Descent of Man and Selection in Relation to Sex.* Indeed, it was the hunter's need to finance slow-maturing children, Darwin thought, that provided the main catalyst for the evolution of our big brains. 'The most able men succeeded best in defending and providing for themselves and their wives and offspring,' he wrote. It was the offspring of hunters with 'greater intellectual vigor and power of invention' who were most likely to survive. According to this logic, males with bigger brains would have been more successful hunters, better providers, and more able to obtain mates and thereby pass their genes to children whose survival was under-written by a better diet. Meat would subsidize the long childhoods needed to develop

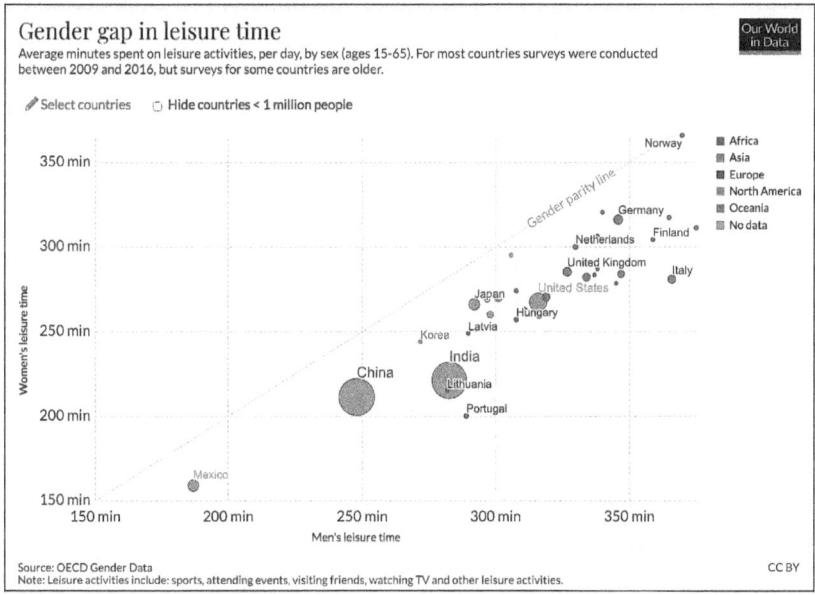

Fig. 1.11. An interesting way to look at how people spend their time globally is through the lens of gender. This figure shows gender disparities in time spent within 33 countries: women spend nearly three times more in unpaid care work compared to men, a whopping total of 1.1 trillion hours each year, which means a lot less leisure time, except in Norway, which is a very rare example.

larger brains, leading eventually to the expansion of brains from the size of an australopithecine's to the size of Darwin's own. Thus did the 'hunting hypothesis' morph into one of the most long-standing and influential models in anthropology... At the heart of the model lay a pact between a hunter who provided for his mate and a mate who repaid him with sexual fidelity so the provider could be certain that children he invested in carried at least half of his genes...

[However] as it became apparent that among foragers (like the Kung) plant foods accounted for slightly more calories than meat, researchers started paying more attention to female contributions...

When Frank Marlowe interviewed Hazda still living by hunting and gathering, he learned that only 36 percent of children had fathers living in the same group...

When anthropologists reviewed a sample of 15 traditional societies, in 8 of them the presence or absence of the father had no apparent effect on the survival of children to age 5, provided other caregivers in addition to the mother were on hand in a position of help.

Living apes cannot be considered to be mainly 'active hunters'. It is rare for chimpanzees to hunt actively, and they only occasionally eat the meat of monkeys. Likewise, gorillas and orangutans almost never eat meat in their natural environments. Clearly eating meat, or hunting, is not crucial for most apes: females and their offspring can live well without it and without the need for males to 'save them with their hunted meat'. Moreover, active hunting has only been common in the last one or two million years of our human history. And, although meat was likely to be important before that time, the humans were mainly scavengers. In a manner similar to hyenas, they ate meat opportunistically rather than by actively hunting and killing animals. This, of course, does not mean that hunting for meat has not been crucial for the sustenance of many nomadic hunter-gatherer societies, since active hunting started. Clearly, it is for societies that depend on eating animals, such as the Nunamiut, who are semi-nomadic people located in Inland, Alaska. But in most nomadic hunter-gatherer societies, hunting actually contributes to less than 50% of the diet, as noted by Coontz and Hrdy, and summarised in Robert L. Kelly's 2013 book, *The Lifeways of Hunter-Gatherers* (see **Fig. 1.12**). As noted by Kelly, in the few groups in which hunting contributes to more than 50% of the diet, sharing meat is usually common, particularly within small nomadic hunter-gatherer units. Thus, one cannot say that if a father dies or leaves the group, the mother and children will die from starvation. In most cases, Hrdy notes, they will still obtain meat from the other hunters. In other words, even in these groups one can say that the dependence is more on the group as a whole, rather than on a specific 'father-the-hunter-savior'.

Of course, as we have seen, apart from sexism, racism and ethnocentrism also play a huge role in science. This is particularly the case with scientific racism, which originated in the 18th century in an interactive way. This takes the form of 'naturalized' so-called 'racial differences' which, of course, do not exist as biologically there are no different 'races' within our species, but which serves to further reinforce, and even exaggerate, societal and political racism. This is a typical cycle resulting from the links between politics, society, science, and the history of science. For example, as Raymond Corbey put in *The Metaphysics of Apes*, the so-called 'lower human races', great apes and, in particular, gorillas 'came to be seen as powerful personifications of wildernesses to be fought heroically and conquered by civilized Westerners.' One of the most influential channels of scientific and cultural discourse, through which the 'beast-in-man stereotype' spread from the 19th century to the 20th, was the psychoanalyst Sigmund Freud. Freud provides an emblematic illustration of the type of scholar that was prominent at the time: the entitled European man who was profoundly racist and misogynistic, and also subscribed to the fictional teleological tales about 'progress' in human evolution. This was

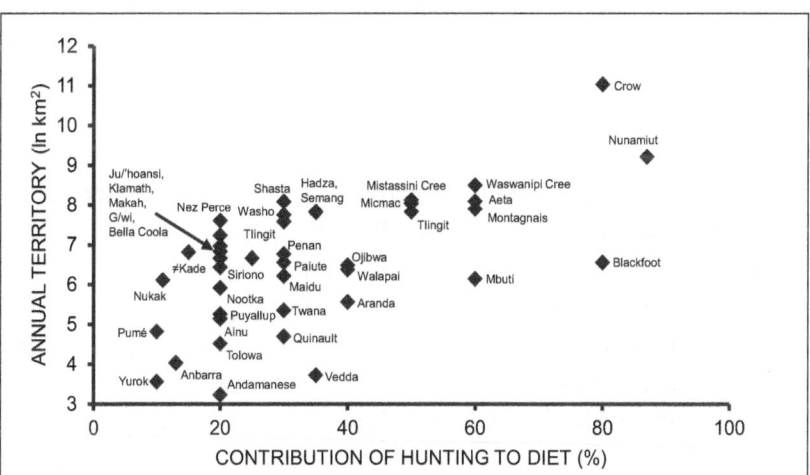

Fig. 1.12. The size of foragers' annual ranges plotted against the percent dependence on hunting; in general, as the dependence on hunting increases so does the size of the exploited territory. In the overall, it is interesting to see how in the vast majority of groups hunting contributes less than 50% to their diet, contrary to the narrative of the 'man-the-hunter-saviour'.

summarised in a short article published in the *Guardian* in 2002, entitled *Scientist or storyteller?*. This piece, by AC Grayling, was written in response to the Penguin publication of the first major translations of Freud's work for over 30 years, which prompted 'serious questions about the nature of Freud's contribution and his legacy'.

As Grayling wrote in his final paragraph regarding Freud's 'psychoanalytic theories': 'Philosophies that capture the imagination never wholly fade. From Animism to Zoroastrianism, every view known to man retains at least a few devotees. There might always be Freudians, and there will always be admirers of Freud's great imaginative and literary powers; these two, as the foregoing remarks suggest, are intimately linked. But as to Freud's claims upon truth, the judgment of time seems to be running against him.'

Among the recent books which state that Freud's theories were not only 'not scientific' but which go as far as to declare him 'a liar and a fraud' and 'putting the final nail in his coffin', I recommend Frederick Crews' book, *Freud – the Making of an Illusion*. This book directly connects these topics with key issues discussed in the present volume: namely, how Freud's huge success and influence were not only due to his use of very appealing, imaginary teleological stories but also to the fact that many scientists, and most lay people, were immersed in such fairytales.

Unfortunately, the use of such fictional ethnocentric and racist narratives in the last two centuries, including terms such as 'apish humans', has led to atrocious consequences at various levels. In American popular culture, a detailed archival content analysis by John Sorenson in his book, *Ape*, reveals that news articles keep on creating implicit associations between 'black criminals' and apes, and that those identified as 'ape-like' are more likely to be executed.

A particularly illustrative and upsetting example of the link between the discovery of non-human primates by Westerners, colonialism and science forms the basis of a colonial propaganda film made in the 1950s in the Belgian Congo. The film was made on behalf of the Belgian government, and circulated widely in Belgian cinemas for an audience consisting of families with children. As described by Corbey: 'the footage shows in great and, by present-day standards, shocking detail how scientists of the Royal Belgian Institute of Natural Sciences shoot and kill an adult female gorilla carrying young. Subsequently the body is skinned and washed in a nearby stream, with the distressed youngster sitting next to it. The adult's skeleton, skin and other body parts were collected for scientific [anatomical] study and conservation, while the live young gorilla was sent to the Antwerp zoo.'

In his book, *Stamped from the Beginning*, Ibram X Kendi discusses how the history of colonialism and the trans-Atlantic slave trade, as well as of American slavery, oppression and discrimination, was deeply rooted in two types of racism. Type B or 'innate' racism was defended by what he calls the 'segregationists', while the more ancient type A or 'cultural-epigenetic' racism was defended by those he designates 'assimilationists':

> The title Stamped from the beginning comes from a speech that Mississippi senator Jefferson Davis gave on the floor of the US Senate on April 12, 1860. This future president of the Confederacy objected to a bill funding Black education in Washington, DC. 'This Government was not founded by negroes nor for negroes', but 'by white men for white men', Davis lectured his colleagues. The bill was based on the false notion of racial equality, he declared. The 'inequality of the white and black races' was 'stamped from the beginning'. It may not be surprising that Jefferson Davis regarded Black people as biologically distinct and inferior to White people—and Black skin as an ugly stamp on the beautiful White canvas of normal human skin—and this Black stamp as a signifier of the Negro's everlasting inferiority. This kind of segregationist thinking is perhaps easier to identify—and easier to condemn—as obviously racist. And yet so many prominent Americans, many of whom we celebrate for their progressive ideas and activism, many of whom had very good intentions, subscribed to assimilationist thinking that also served up racist beliefs about Black inferiority. We have remembered assimilationists' glorious struggle against racial discrimination, and tucked away their inglorious partial blaming of inferior Black behavior for racial disparities. In embracing biological racial equality, assimilationists point to environment—hot climates, discrimination, culture, and poverty—as the creators of inferior Black behaviors. For solutions, they maintain that the ugly Black stamp can be erased—that inferior Black behaviors can be developed, given the proper environment. As such, assimilationists constantly encourage Black adoption of White cultural traits and/or physical ideals...
>
> ... Historically, there have been three sides to this heated argument. A group we can call *segregationists* has blamed Black people

themselves for the racial disparities. A group we can call *antiracists* has pointed to racial discrimination. A group we can call *assimilationists* has tried to argue for both, saying that Black people *and* racial discrimination were to blame for racial disparities.

Kendi also provides a powerful example of the link between racist stereotypes and scientific biases, which resonates with us following the Covid-19 pandemic and the importance of vaccines to try and reduce its impact:

On April 21, 1721, the HMS Seahorse sailed into Boston Harbor from Barbados. A month later, Cotton Mather logged in his journal, 'the grievous calamity of the smallpox has now entered the town.' One thousand Bostonians, nearly 10 percent of the town, fled to the countryside to escape the judgment of the Almighty.

Fifteen years prior, Mather had asked Onesimus one of the standard questions that Boston slaveholders asked new house slaves: Have you had smallpox? 'Yes and no,' Onesimus answered. He explained how in Africa before his enslavement, a tiny amount of pus from a smallpox victim had been scraped into his skin with a thorn, following a practice hundreds of years old that resulted in building up healthy recipients' immunities to the disease. This form of inoculation—a precursor to modern vaccination—was an innovative practice that prevented untold numbers of deaths in West Africa and on disease-ridden slave ships to ports throughout the Atlantic. Racist European scientists at first refused to recognize that African physicians could have made such advances. Indeed, it would take several decades and many more deaths before British physician Edward Jenner, the so-called father of immunology, validated inoculation.

Cotton Mather, however, became an early believer when he read an essay on inoculation in the Royal Society's *Philosophical Transactions* in 1714. He then interviewed Africans around Boston to be sure. Sharing their inoculation stories, they gave him a window into the intellectual culture of West Africa. He had trouble grasping it, instead complaining about how 'brokenly and blunderingly and like Idiots they tell the Story'...

... The only doctor who responded to Mather was Zabadiel Boylston, President John Adams's great-uncle. When Boylston announced his successful inoculation of his six-year-old son and two enslaved Africans on July 15, 1721, area doctors and councilmen were horrified. It made no sense that people should inject themselves with a disease to save themselves from the disease. Boston's only holder of a medical degree, a physician pressing to maintain his professional legitimacy, fanned the city's flames of fear. Dr. William Douglass concocted a conspiracy theory, saying there was a grand plot afoot among African people, who had agreed to kill their masters by convincing them to be inoculated. 'There is not a Race of Men on Earth more False Liars than Africans,' Douglass barked.

Politically tribalistic narratives of 'us' versus 'them' have also played a significant role in the history of science. A particularly sad example of how science is not as 'objective' as it is often portrayed, can be seen in the case of Nikolai Vavilov. Vavilov was one of the brightest Soviet biologists, and many of his evolutionary ideas are now being acknowledged, as I can attest to when I gave a talk some years ago at the Russian Academy of Sciences in Moscow. For decades, Vavilov was actively persecuted in the Soviet Union and his work almost forgotten, while that of Trofim Lysenko—who was basically a charlatan—was increasingly seen as of scientific genius. Worse than that, there is strong evidence that Lysenko, alongside Stalin, was directly involved in the persecution of Vavilov, which resulted in his imprisonment in a miserable Soviet prison, and death due to disease and starvation. This is how far scientific biases, and political teleological narratives, such as those invoked in communism and Marxism, as well as capitalism and all types of political storytelling, can go.

In this particular case, the problem for Vavilov was that his ideas about genes and their role in evolution and plant breeding were opposed to Lysenko's 'law of the life of species' and Stalin's communism which regarded genetics as a field connected to the West and capitalism. A book that I strongly recommend about Vavilov, Lysenko and Stalin, is Peter Pringle's book, *The Murder of Nikolai Vavilov*.

Briefly explained, biology epigenetics is the study of changes, including heritable ones, that do not involve alterations in the DNA sequence. Of course, we now know that epigenetics plays a much more important role than most Neo-Darwinists were defending at that time, so in this respect some of the biological ideas defended in Stalin's communist narratives were not all wrong. However, the main political narrative was about much more than this: namely, that the 'real theory of the knowledge of the world was given by Marx, Engels, and Lenin', which is obviously wrong. And whether or not some of Vavilov's ideas about genetics might have eventually been proven wrong, his personal persecution, imprisonment, and ultimately death in prison do highlight in a very tragic way the importance of politics in science. As explained by Pringle:

> When Stalin praised Lysenko in a Kremlin hall packed with plant breeders and collective farm workers, he marked a new and destructive era in the history of Soviet biology. A jubilant Lysenko quickly claimed a series of new plant breeding achievements, each one less credible than the last, yet each one carefully couched as 'practical science', and linked expressly to Stalin's demands for quicker results. The agricultural bosses followed their leader by giving Lysenkoism renewed support. Commissar Yakovlev praised Lysenko as a 'practical worker whose vernalization of plants has opened a new chapter in agricultural science'... a scientist who was now 'heeded by the entire agricultural world, not only here, but abroad as well'. Lysenko's 'people' would become the 'backbone of the real Bolshevik apparatus'. Vavilov did not respond, at first. It was not in his nature to engage in an academic brawl, and, even if he had wanted to, the stakes were now much higher. For a geneticist to criticize Lysenko's claims when Soviet agricultural production was failing was to seem unpatriotic at best and, at worst, to be engaging in economic sabotage. Lysenko himself encouraged such treasonable thoughts. His speeches were routinely laced with references to the nefarious activities of 'bourgeois scientists', 'saboteurs', and 'class enemies'. In the Kremlin speech to the collective workers, the passage that drew the most applause, and Stalin's extraordinary exclamation, was not about any new achievement in crop production but Lysenko's attack on bourgeois science. All the bourgeois academics ever did was 'observe and explain phenomena,' he said, while socialist science aimed to 'alter the plant and animal world in favor of the building of socialist society.'
>
> [Many years later] on January 24, 1943, emaciated and suffering from a fever, he [Nikolai Vavilov] was moved to the prison hospital. As he entered he introduced himself, 'You see before you, talking of the past, the Academician Vavilov, but now according to the opinion of the investigators, nothing but dung.' In the hospital he com-

plained of chest pains and shortness of breath. He had diarrhea and was hardly eating. He was put on 'meals of the 2nd type' that included milk, and glass jars were placed on his chest in an effort to extract the fever that the doctor noted might be a recurrence of malaria. A committee of Saratov jail doctors examined Nikolai Ivanovich. They noted that he was complaining of overall weakness. And they saw 'maceration, pale skin, and swelling in the feet'. Their diagnosis was: 'dystrophy from prolonged malnutrition'. On January 26, 1943, at seven o'clock in the morning, Nikolai Ivanovich's heart stopped beating. The official cause of death was recorded as pneumonia, a cold caught in the prison exercise yard, perhaps. This eminent plant hunter who had a plan to feed the world had died of starvation.

For those readers interested to know more about the topics discussed both in this section and in Chapter 1, Rui Diogo has published a book analysing them in a broader, multidisciplinary way. It is entitled: *Meaning of Life, Human Nature, and Delusions – How Tales About Love, Sex, Races, Gods and Progress Affect our lives and Earth's splendor.*

CHAPTER 2

UNTOLD STORIES: MUSLIM SCHOLARS IN ANATOMICAL AND MEDICAL SCIENCES

"Therefore, in medicine we ought to know the causes of sickness and health."
(Avicenna)

Seeing Human Anatomy in the Body of Macaques

Scholars in Ancient Greece incorporated medical and clinical anatomy theories from the Egyptians into their own studies. This paved the way for the development of anatomical sciences (Loukas *et al.*, 2011; Standring, 2016). As noted above, Galen (**Fig. 1.1**), a Greek living in Rome during the time of the Roman Empire, was, for several centuries, considered the 'father of human anatomy'. This is still the case in popular culture. However, many scholars and historians of science now recognise that this reputation might be unjustified, as it is unlikely that Galen ever dissected human bodies, or at least, only dissected a few. In fact, within the Greco-Roman culture of the dead body, the human corpse was considered impure (Park, 2006). Most of Galen's descriptions of the 'human body' were actually based on dissections of animals, such as sheep, oxen, pigs, dogs, bears, and, particularly, the 'Barbary ape'; an Old World monkey (*Macaca sylvanus*: **Fig. 2.1**), which has a greatly reduced vestigial tail and thus, su-

Fig. 2.1. Mother and baby of the monkey species *Macaca sylvanus*, or 'barbary ape': for more than a millennium scholars and health practitioners accepted Galen's descriptions of the hard and soft tissues of various non-human animals—including, and particularly, of this monkey species—as the unquestionable 'truth' about the anatomy of the human body.

perficially, resembles an ape (which has no tail) (Singer, 1957, 1959; Cole, 1975).

These studies explain the odd structures described by Galen as features of the 'human body'. The accounts often matched certain anatomical features of the Barbary monkey, which, in fact, has marked differences when compared to humans, particularly concerning soft tissues such as muscles (Diogo and Wood, 2012). For instance, Galen's description of the 'human body' did not refer to two muscles that are almost exclusively present—within the order of Primates—in humans, namely the muscles of the forelimb: i.e., the flexor pollicis longus and extensor pollicis brevis. Instead, Galen described the forelimb muscles typically present in monkeys, such as *Macaca* (more examples given in **Table 2.1**). Moreover, he inaccurately described certain features that are similar in both humans and macaques. In doing so, he failed to recognise that the extensor carpi radialis brevis and longus are separate, and distinct, forelimb muscles in both humans and macaques (**Table 2.1**).

Table 2.1: Muscles' origins, insertions, and functions according to Galen, who was mainly based on dissections of Old World Monkeys. Information and footnote in this table are extracted from Singer (1956).

Muscles on the inner side (flexor surface) of forearm

Galen's Description	Modern name	Origin	Insertion	Function
1) The first muscle is on the surface of the mid forearm, under the skin, between ulna and radius. It ends in a flat tendon and extends under the smooth hairless part of the hand	Palmaris longus			
2) On either side of the ligaments (retinacula) on the inner side (of the arm) is a muscle flexing the wrist. It is in a line with the little finger	Flexor carpi ulnaris	Articulation of elbow	Onto straight and cartilaginous bone at wrist which is in line with little finger (pisiform)	Bends wrist
3) Other with the index	Flexor carpi radialis		Plunges deep immediately after articulation and reaches base of metacarpal of index, to which it is attached	Bends wrist
4) Produces four tendons	Flexor digitorum superficialis	Springs from inner condyle of humerus and is connected for short space with ulna	Four tendons; each splits into two. They inserts onto all digits except thumb, at beginning of second phalanx	Flexes second joint
5) The other tendon/head lying beneath the former and splits into five parts. Each reaching the last joint of the digit, inserting there	Flexor digitorum profundus + Flexor pollicis longus	Attached to both bones. Forward outgrowth of ulna (coronoid process) in elbow region	By five parts into last joint of five digits without splitting; the one that goes to the thumb passes on to second joint	
A muscle descending from above the radius	Brachioradialis		Does not end below in tendon like those so far mentioned	

MUSLIM SCHOLARS IN ANATOMICAL AND MEDICAL SCIENCES   37

Muscles on the outer side (extensor surface) of forearm

Galen's Description	Modern name	Origin	Insertion	Function
1) Placed in the midst of the others	Extensor digitorum	Outer condyle of humerus		Moves digits (other than thumb)
2) Secondly, the tendon/head which draws the two little fingers to the side. This last is divided into two tendons.	Extensores digitorum proprii IV et V (Extensor digiti minimi to digits 4&5)	Outer condyle of humerus		Draws two little fingers to lower part of hand, i.e., to side away from others
3) Then you must raise the remaining one, the third, which initiates a like movement in the two[1] bigger fingers	Extensores digitorum proprii II et III (Extensor indicis to digits 2&3)	From greater part of length of ulna		Initiates a similar movement in two bigger fingers
4)	Extensor pollicis longus	Upper end of ulna		Moves thumb
5) The muscle that extends the wrist at the little finger	Extensor carpi ulnaris	Outer condyle of humerus	By single tendon at back off fifth meta-carpal[2]	
6) & 7) At the thumb region, a strong ligament binds the head of two tendons	Abductor pollicis longus + extensor pollicis brevis	Occupies whole depth between radius and ulna	One tendon inserted into metacarpal of thumb (on sesamoid there) and other into thumb itself immediately after first articulation	
8) The remaining muscle on the outer side of the forearm	Extensor carpi radialis longus + brevis	Highest part of outer condyle and reaching point on humerus above it	By double tendon onto second and third metacarpals, index and middle fingers	Extends wrist

Wrist muscles

Galen's Description	Modern name	Origin	Insertion	Function
Under the membrane a certain small muscle runs transversely and extending from ulna to radius	Pronator quadratus			
	Supinator	Origin above this (radius)	Continued and united with origin of extensor carpi radialis	Turns palm upward
Between the two heads (flexor carpi ulnaris and flexor carpi radialis) lies the origin of the muscle that runs down into the skin of the hand	Palmaris longus (see above)			

There are four muscles; by these the whole hand is supinated or the reverse

Galen's Description	Modern name	Origin	Insertion	Function
A) Two at the wrist: 1) Transverse muscle that lies between radius and ulna.	Pronator quadratus	Ulna	Reaches radius	Turns radius inwards and initiates prone position
2) The yet longer and more fleshy muscle, which lies altogether above this	Extensor carpi radialis longus + brevis (see above)	See above	See above	Moves hand to supine position
B) Two are seen to reach the upper part of the forearm and move the upper part of the radius	Pronator teres	From (inner) condyle of humerus		
2) The other on the outside and is smaller	Supinator (see above)		Radius	

Small muscles of the hand

Galen's Description	Modern name	Origin	Insertion	Function
Small muscles beside the tendons, they are four in number for the fifth, fourth, third, and second digits	Lumbricales	Four sheaths surrounding tendons (of flexor digitorum profundus)	Reach sides of fingers and producing very slender tendons	Flex third joint
The thumb is moved by two other muscles	Abductor pollicis brevis	First bone at wrist (navicular)		Drawing it away from other fingers
	Adductor pollicis	Third metacarpal		Drawing it towards index
The little finger drawn away by a muscle	Abductor minimi digiti	Bone of carpus corresponding to it (pisiform)		Drawn little finger away
There are two for each finger	Palmar interossei? Or, more likely, flexores breves profundi, as seen in Old World Monkeys		Reach first articulation on inner side and attached to sides (of phalanx)	Bends first joint slightly
All the others	Dorsal interossei? Or, more likely, intermetacarpales, as seen in Old World Monkeys	Ligament at wrist and metacarpus at roughly same articulation of bones (as palmar interossei)		
Those belonging to the thumb	Two heads of flexor pollicis brevis	Ligament that confines two tendons of muscles which flex fingers (flexor retinaculum)		

Arm muscles

Galen's Description	Modern name	Origin	Insertion	Function
The anterior muscle of the arm has two heads	Biceps brachii	One attached to ridge on neck of shoulder blade and other to process which some call "like an anchor" while others "like a crow's beak"	Attached to radius by strong tendon	Flexes joint and bend it slightly inwards
Beneath the previous muscle, which also encircles the humerus	Brachialis	From two fleshy heads: one at back of humerus and other more to front	Attached to ulna	Flexes joint and bends it slightly outwards
Three others united: two muscles which extend the forearm	Long head of Triceps brachii	Lower side of scapula halfway down upper part	These coalesce in upper arm and inserted in crook of ulna (olecranon) by flat tendon	Stretching each part of muscle lead to extension of forearm with inclination to sides
	Lateral head of Triceps brachii	Back of upper humerus below its head		
Another muscle lies under it and surrounding the bone of the upper arm obliquely	Medial head of Triceps brachii		Unites with second muscle and thought to be a part of it by anatomists as indeed it is if you think of this muscle as single	Straight and direct extension at elbow joint

Muscles of the lower limb

Galen's Description	Modern name	Origin	Insertion	Function
On the surface under the skin appears a flat tendon	Sartorius	Upper end of it has fleshy origin from middle of ridge (anterior superior spine) of ilium	Inserted in tibia below knee	Brings leg into position that boys use in palaestra in (changing legs) when they throw other leg on thigh
Beyond the point where this (Sartorius) muscle becomes tendinous is another insertion	Gracilis	Front of pubic bone	Tendon passing into tibia	Raises and rotates leg inward
In the same conjunction on the inner side of the tibia, a third attachment of a flat tendon	Semimembranosus proprius	Outer and lower part of ischium	Inner side of tibia	Rolling leg backward
The fourth of those that descend to the tibia	Biceps femoris	Lies at farthest point outside ischial bone	Descend to tibia on outside	Draws whole leg outward with simple motion
There is (in the hip region) a fifth muscle	Accessory Semimembranosus	ischial bone	Does not reach tibia like first three but goes to lower head of femur and to origin of muscles of leg on inner side	
In contact with the head of the calf muscle	Semitendinosus			
	Medial head of Gastrocnemius	Thigh condyle		Turning tibia backward and somewhat inward on thigh

Muscles moving the knee-joint

Galen's Description	Modern name	Origin	Insertion	Function
There are four of them: 1) The highest	Rectus femoris	Ridge of ilium	Toward knee-cap	Extend knee
2) Next is one much larger, lower down, and on the outer part of the thigh toward the buttock	Vastus lateralis	Great trochanter and neck of femur		
3) Runs from about the middle of the thigh to its lower end	Vastus intermedius		Runs down straight through front parts of thigh as far as patella and remaining entirely fleshy	
4) Another tendon also reaches the same place as the muscle (rectus femoris)	Vastus medialis	Anterior region of femur	Ends by inner side of thigh and acquiring more membranous end	
Large muscle, which occupies practically the whole postero-medial part of the thigh,	Adductor magnus		Reaches parts round knee-joint	Draws leg after it little if at all

Muscles of the Hip

Galen's Description	Modern name	Origin	Insertion	Function
	Biceps femoris/Semi-membranosus?	Fibers from back of femur	Pass up toward ischium	Bends hip joint
	Adductor longus	Fibers on inner side	Reach inner part of pubic bone	
The underlying muscle which occupies the great foramen	Obturator externus		Passes into tendon	
Small livid muscle	Pectineus	Deeper parts of pubic bone	United to lower part of small trochanter by tendon	Initiates oblique motion in thigh towards inside
Another muscle produces a yet stronger tendon of attachment and it is the only muscle in the loin region that is three-fold	Iliopsoas	Both from ilium and from both loin-muscles	To remaining part of small trochanter	Flexes and rotates thigh inward
Set on the surface under the skin	Tensor fasciae latae	Whole straight ridge of ilium		
Fleshy, passes toward the hip-joint and is continuous with it (gluteal fascia)	Gluteus maximus			Extends thigh with slight outward[3] inclination
When this muscle has been removed, there remains another	Gluteus medius	Whole back of ilium and embraces to some extent also neighboring bones (sacrum)	Its tendon attaches to apex of great trochanter and extends even in front	
Small muscle that you may think this to be part of the "large muscle" (gluteus medius)	Gluteus minimus	Outer and lower parts of ilium	Great trochanter	
Another muscle dark in color is hidden there under the "large muscle" (gluteus medius)	Piriformis	Inner lateral parts of sacrum	Great trochanter	Rotates head of thigh
Next there are two others that are completely hidden	Obturator externus and obturator internus? Or include gemellus inferior and/or inferior, as obturator was already described above	Pubic bone and occupy (obturator) foramen	By strong tendons in hollow (digital fossa) by large trochanter	Rotates head of femur outwards

Muscles of the Leg

Galen's Description	Modern name	Origin	Insertion	Function
Two muscle-heads	Gastrocnemius (see also above)	Back of femur at roots of condyles	Heads join and become one tendon (tendocalcaneus) which lies beneath and adjoins muscle (plantaris) and inserts onto end of heel at back	Pulls heel that way
At this point (where the two heads join) a considerable strand splits off from the outer head. This becomes a muscle	Plantaris		Ending gradually in flat aponeurosis under sole	
An attachment higher up, belonging to another muscle	Soleus	Fibula[4]	Reaches heel	
Other muscles continuous with them to the side which reach to the underside of the foot	Flexor digitorum fibularis & tibialis	Heads are between calcaneum and tibia and one (flexor digitorum fibularis) placed at lower end of talus where it lies beside calcaneum	Not always distributed in same way	One moves toe corresponding index and little toe and other moves middle and fourth toes Both united in a common tendon and move big toe
	Tibialis posterior	End of tibia	Fuses with first bone of tarsus on inner side (navicular)	Bends whole foot backwards
First muscle on the outer side of the leg	Fibularis longus	Extending along fibula to its upper head	Head of first bone (metatarsal) of big toe	Bends it (first bone) at articulation
Second muscle lying beside this (peroneus longus) and thought to be part of it	Flexor digitorum fibularis or Flexor hallucis longus	Has common head above and lies along it (fibularis longus) on outside throughout leg	Head of first phalanx of big toe (and into digits III and IV)	
Another thin muscle	Tibialis anterior	Between fibula and tibia	Side part of big toe	
Tendon-head lying under the (transverse) ligament	Extensor digitorum longus			Four tendons which extend four toes
Over this muscle (the previous) you will see another	Extensor hallucis longus	Fibula	At end of tarsus in great toe above inner side	
	Fibularis longus [5]	See above	See above	See above
	Quadratus plantae	From delicate ligaments attached to portions of flesh		
	Abductor digiti minimi	Where head of above mentioned muscle ends	Ends in round tendon	Drawing little toe outwards
	Tibialis posterior?	From remaining part of heel	Fuses with first bone of tarsus on inner side (navicular)	Turns whole foot upwards

The Five Kinds of Muscles of the Mouth

Galen's Description	Modern name	Origin	Insertion	Function
	Platysma	From all neck vertebrae, scapulae and clavicles	At lips and some passing from left to right of lips while others reverse way	Opens mouth on either side towards neck
	Temporalis			Draws jaw up in biting on anything or in nibbling or in shutting mouth
Lying on the jaw. Each masseter forms two muscles and each comes from its own head to a common end	Masseters; One of heads you will see in cheek, other lies along whole jugal bone and is not at all sinewy		In lower jaw	Moving jaw either way first: draws up jaw slightly to front Second: opposite movement
	Nasolabiales		Conjoined and fused with upper lip (through leuvtores labii)	Moving alae of nose
Lie under the skin of the forehead	Frontalis	See below	See below	See below
The muscle hidden within the mouth, which is contiguous to them (masseter and temporal)	Pterygoid	From wing-like (pterygoid) outgrowth of skull	Flat parts of lower jaw where there is place made slightly hollow to afford access to muscle	
Muscle attached to the flat part of the lower jaw	Buccinator			

Muscles of Forehead and Neck, and Movers of the Head

Galen's Description	Modern name	Origin	Insertion	Function
A single muscle running on either side of the spine	Trapezius	Bone of skull at nape	Inserted into sharp ridge on scapula	Draws up shoulder to head
They have a transverse origin and they extend along the spinous processes through the neck. When they approach the scapula, cohere with the muscles lying beside it on either side	Rhomboid complex, including Rhomboideus occipitalis	Middle region of skull at inion (external occipital protuberance)		Draw up base of scapula towards inion (external occipital protuberance); extend head with trapezius by fleshy projections throughout neck
	Spinal muscles	From each of upper cervical vertebrae through powerful ligaments	Attached to neighboring vertebrae	
Flat pair and their fibers are oblique	Splenius	Skull at inion (external occipital protuberance) transversely like those above mentioned	To spine	
In the contrary direction to these, the fibers of the muscles under them run slanting forward towards the transverse processes of the vertebrae	Obliquus capitis superior, obliquus capitis inferior, and rectus capitis posterior major	Inion (external occipital protuberance)	One extends to spines of vertebrae Second to their transverse processes Third in between	Draw skull back
Small muscle that fastens the first vertebra to the head	Rectus capitis posterior minor	Axis	Skull at inion near its middle	
	Obliquus capitis superior	Skull	Transverse process of first vertebra	
The third pair	Rectus capitis posterior major			Initiates sideways movement of head along line of its fibers
A fourth pair at an angle to the third	Obliquus capitis inferior	First vertebra	Second vertebra	

MUSLIM SCHOLARS IN ANATOMICAL AND MEDICAL SCIENCES

Muscles uniting the Skull with Sternum and Clavicle

Galen's Description	Modern name	Origin	Insertion	Function
From two starting points, one lying behind the ear and the other under it	Cleidomastoid and Sternomastoid heads of Sternocleidomastoideus	One springing from back parts of ear at root of ear and other reaching ear along transverse line	Coming down to end of collar-bone at sternum	Either move sternum or clavicle with thorax towards side of head or advance head

Muscles which move the Scapula

Galen's Description	Modern name	Origin	Insertion	Function
There are two by the spine	Rhomhoideus major and rhomboideus minor			Draw scapula backwards
Long and thin, which fastens the scapula to the bone called hyoid at the beginning of the larynx	Omohyoideus	From parts at larynx	To scapula	Pulling it towards front of neck
	Longissimus capitis	Spinous processes of second vertebra	To third vertebra	
When you have examined the five (vertebrae) of the neck, you will find a superficial muscle near the thoracic inlet	Levator scapulae	Five vertebrae of neck and seven vertebrae of thorax		
One moves the scapula	Serratus anterior	Subcostal arch	Inserted into middle parts of ribs at their maximum convexity	Dragging scapula downwards

The Twin Muscles that open the Mouth

Galen's Description	Modern name	Origin	Insertion	Function
Those opening the jaw	Digastricus anterior + Digastricus posterior	Stone-like (petrous) bone of skull	Extend up to very end of jaw (the chin) so muscles of two sides meet	

Muscles uniting Thorax to Humerus and Scapula

Galen's Description	Modern name	Origin	Insertion	Function
Large muscle, it is two fold, its fibers crossing each other like the letter X	Pectoralis major	Whole sternum and nipple lying on it	Shoulder joint	Fibers from higher part of sternum: brings humerus to thorax without pulling it downward Other fibers: gives it an oblique downward movement
	Pectoralis minor	Sternum, with ribs 2 to 6		Adducts humerus
In the sides of the thorax are two muscles coming up from below: one is thin on the surface	Latissimus dorsi	Lumbar vertebrae		
The other muscle comes from below	Lower part of Trapezius	Spines of vertebrae especially those of false ribs		

Shoulder Muscles

Galen's Description	Modern name	Origin	Insertion	Function
	Subclavius	Coming down from clavicle	To first rib	
Muscle of the shoulder that is continuous and united with the largest of the muscles from the sternum (pectoralis major)	Deltoid	Clavicle and scapula	Below on humerus	
	Teres major	Side of scapula		
Another muscle which runs upon the humerus	Teres minor		Front of humerus below articulation	
Two muscles extended along the scapula one above and the other below (the spine of the scapula)	Supraspinatus and Infraspinatus			Move humerus obliquely one outward toward clavicle and other inward towards lower part of scapula

Muscles moving the Thorax

Galen's Description	Modern name	Origin	Insertion	Function
One lying close to the first rib	Sternocostalis (Transversus thoracis)			
	Serratus anterior	See above	See above	Draw up whole thorax except lower part which is moved by diaphragm
On either side of it [serratus anterior] lie other muscles. One is in the front of the thorax and the other in the back	Scalenus longus	Second (cervical) vertebra	In first five ribs by strong ligaments	Both drawing up ribs
	Serratus posterior superior	Last three cervical and first thoracic vertebrae	Each of them having membranous ligament as its head interwoven with spinal muscles	
Another (pair) belonging to the first ribs	Scalenus brevis anterior			Drawn first ribs up and and dilate upper part of thorax
Two other pairs of muscles along the length of the thorax, one is fleshy tissue pertaining to the spine and the other is membranous tissue to the sternum.	Iliocostalis dorsi spinalis,	Begins and ends with thorax	Lower end inserted into spinal muscle	Protects and pulls in ribs
	Thoracic part of Rectus abdominis	Tendinous intersection	Cartilages of false ribs in left and right	
Another pair of muscles outside the thorax	Serratus posterior inferior		Inserted along last ribs	Draws down last ribs of thorax
The fibers are clearly seen to be oblique in the mid-part of the ribs				
One running obliquely from below upward and the other in a reverse direction, crossing like an X	Intercostal muscles			
	Diaphragm	In numerous delicate ligaments springing from bones with simple flesh coagulated round them		

MUSLIM SCHOLARS IN ANATOMICAL AND MEDICAL SCIENCES   45

The Abdominal Muscles

Galen's Description	Modern name	Origin	Insertion	Function
The largest and outermost of all	Abdominal external oblique	First of attachments situated by sixth rib then it springs from all other ribs	Spreads upon abdominal muscles then inserts in innominate bone at pubes	
Whose fibers have an oblique position at right angles to the first	Abdominal internal oblique	Bone of flank (ilium)	Inserted into ends of four false ribs	
Under them is the 'third' stretched out lengthwise	Rectus abdominis			
The remaining fourth pair	Transversus abdominis	Straight line of ilia and transverse processes of vertebrae in loins	Attached to peritoneum	

The Lumbar Muscles

Galen's Description	Modern name	Origin	Insertion	Function
Under the esophagus	Longus colli			Bend upper part of spine
Single muscles, one on either side along the spine. They are united at their origin above throughout the loin. When they approach the broad bone (the sacrum), they separate	Psoas muscles: Psoas minor is ligament	Higher portion of fleshy lumbar mass and advances through inner region	Where pubic bone joins ilium	Bend spine at loins
	Psoas major	Latter arises lower beside it and on outside and (lower still) that which comes to it from ilium	Descends to small trochanter of femur	

The Intrinsic Spinal Muscles

Galen's Description	Modern name	Origin	Insertion	Function
Their fibers are oblique. Some running from the spine forwards and downwards and some the opposite way	Longissimus infermedius and/or Iliocostalis	Second cervical vertebra	Each having two separate heads	
Muscles spring which run up gently slantwise to the last ribs of the thorax	Iliocostalis lumborum		Second and third of false ribs	Draw down last ribs

Modern day anatomists are well aware of Galen's inaccuracies when it came to his descriptions of 'the human skeletal system', which was also clearly based on observations of other animals. For example, he described the left and right lower jaws, which in humans are fused into a single jaw bone, a separate premaxilla; in humans this bone is fused with the maxilla, and up to seven distinct sternal segments—in humans there are usually three. Galen's inaccuracies about the 'human body' also applied to other body systems, as well as functional morphology and physiology. For instance, he stated that air enters the heart directly from the lungs, and that blood passes from one side of the heart to the other through small holes in the septum dividing the right and left ventricles. These errors had crucial repercussions for anatomy, in particular, and biology and science in general, because Galen impressed both the people of his time and successive generations such that, for centuries, his work was regarded as almost infallible (Singer, 1957, 1959).

With regards to this book, it can be seen that Galen's work is subject to great scientific bias. For more than a millennium, scholars and physicians have, dogmatically, accepted Galen's descriptions about the 'human body', despite the fact that they were, in reality, based on observations of other animals. Not only that, but for more than one millennium, scholars and physicians in the West—and also in the Muslim world where Galen's works were very influential—knew more about the anatomy of macaques than of their own species. In fact, the quasi-religious way in which Galen's work on 'human anatomy' was viewed is in great part related to his belief in a single God (although he remained a pagan). He defended the idea that every organ in the human body had been created by God in the best possible form and for its perfect use; a concept that fitted well with monotheistic religions, such as Christianity (Cunningham, 1997) and Islam. Paradoxically, because the anatomy of macaques is similar to that of humans, Galen paved the way for an approximated understanding of the anatomy of various structures of the human body such that Vesalius, in 1543, used the Galenic *Corpus* for the basis of his own *De Humani Corporis Fabrica Libri Septem* (**Fig. 1.3**). That is, for those that were able to escape from the quasi-religious idolisation of Galen and/or the belief that he was infallible, as Vesalius and the Muslim scholars listed in **Fig. 2.2** did, Galen's work provided a helpful basis that was later corrected. Indeed, historians often view Vesalius' works (p.2) as a 'corrected and expanded version' of Galen's writings. As Cole (1975: 42), put it, the 'tendency to see nothing in Galen but his errors reveals a lack of knowledge and understanding, and is just as wrong as was the servile faith which for centuries proclaimed his infallibility.'

850	950	1050	1150	1250	1350	1450
	Al-Razi (Rhazes) (865-925)					
		Al-Akhawayni Bukhari (Joveini) (?-983)				
		Ali Abbas (Haly/Hali Abbas) (930-994)				
			Ibn Sina (Avicenna) (980-1037)			
		Ibn Al-Haytham (Alhazen) (965-1040)				
				Ibn Rushd (Averroes) (1126-1198)		
				Al-Baghdadi (1162-1231)		
					Ibn Al-Nafis (1210-1288)	
					Mansur ibn Ilyas (1380-1422)	
850	950	1050	1150	1250	1350	1450
235	338	441	544	647	750	853

Fig. 2.2. Timeline of Muslim scholars from the 7th to the 13th centuries. The lower row shows the years in Hijri, that is, according to the Islamic calendar.

What is particularly relevant for the present book is the idea, promoted in Western literature, that 'after Galen we encounter no biological activity for centuries', and that the 'revival of anatomy' began predominantly with Vesalius (1514–1564) (Singer, 1959: 63–64, 98). However, some textbooks recognise that from around the 7th or 8th centuries until the 15th to 16th centuries there was a Golden Islamic Age, during which knowledge and science flourished (Hehmeyer and Khan, 2007). Despite this, many textbooks argue that the primary 'function' of Muslim scholars—particularly concerning anatomy and biology—was to 'save' Greek knowledge from total destruction by translating many scientific books, including numerous medical books, rather than discovering new facts (e.g., Persaud, 1984; Muazzam and Muazzam, 1989). The translation process was, indeed, very active during the 8th century, being patronised by the Caliph (Ruler) Harun Al-Rashid (786–809), who designed *Bayt Al-Hikma* (*The House of Wisdom*) in Baghdad. Arab, Persian and Christian scholars like Hunayn ibn Ishaq who translated more than 129 works of Galen, were part of this academic institute (Meyerhof, 1926; Savage-Smith, 1995).

In recent decades, many historians have come to recognise that the traditional 19th century view of history, whereby the European 'Middle Ages'—the period between the demise of Rome and the Renaissance—completely lacked scientific innovation, is far from accurate. Numerous science historians have adopted a broader outlook, recognising that the various forms of exchange and circulation of knowledge between different geographical and cultural regions, and Arabic speakers, as well as the Persians, have made major contributions to biological knowledge (Newman, 1998; Roger, 2000; Syed, 2002; Ihsanoglu, 2004; Tibi, 2006; Pormann and Savage-Smith, 2007; Abdel-Halim, 2008; Russel, 2010; Gotthard, 2012; Jurgen, 2012; Campbell, 2013; Yarmohammadi *et al.*, 2013a, b; Dalfardi *et al.*, 2014a, b, c; Ziaee, 2014).

However, the contribution of Muslim scholars to anatomy continues to be an untold story, not only with the public but even amongst most anatomists. For instance, major works written a few years ago (Persaud *et al.*, 2014) about Muslim contribution to science mainly focus on medicine, and the few focusing on anatomy are often specific and/or published in specialised journals and either refer to a specific Muslim scholar or a certain period of time and/or region *e.g.*, Persia (Shoja and Tubbs, 2007). To the best of our knowledge, before the present work, there was not a single, accessible publication reviewing those Muslim scholars who, prior to Vesalius, had made significant anatomical discoveries.

In order to address the question: *prior to Vesalius, were Muslim scholars able to build and improve upon anatomical knowledge using translations of Greek books as a foundation,* we undertook a detailed literature review that included: 1) works published in various languages and spanning several centuries; 2) a detailed analysis of the original texts of Muslim scholars who, prior to Vesalius, had published works on human anatomy; and 3) a detailed comparison between the anatomical descriptions of the human body by these Muslim scholars and 'human anatomy' as described by Galen, including specific nerves (**Fig. 2.9**) and muscles (**Tables 2.1–2.3**): see below.

Major pre-Vesalius Muslim Scholars who Studied Human Anatomy

Based on an extensive literature review and anatomical comparisons, we list nine key pre-Vesalius Muslim scholars (**Fig. 2.2**). We reviewed and translated the original manuscripts of Al-Razi [*Al-Hawi fi Al-Tibb* (*Comprehensive*) and *Ketab Al-Mansuri*], Ibn Abbas (*The Complete Art of Medicine*), Ibn Sina, or Avicenna (**Fig. 1.2**) (*The Canon of Medicine*), Ibn Al-Haytham (*Book of Optics*), Ibn Rushd [Al-Kulliyat (*Generalities*)], Al-Baghdadi [*Al-Tibb min Al-Kitab wa-Al-Sunnah* (*Medicine from the Holy Book and the Life of the Prophet*) and *Al-Ifada wa'l-I'tibar*], Ibn Al-Nafis [*Mujaz Al-Qanun* (*Commentary on Anatomy in Avicenna's Canon*)], and Mansur ibn Ilyas (*Mansūr's Anatomy*). The texts of these scholars were written in Arabic, with the exception of those by Mansur ibn Ilyas and Al-Akhawayni, which were originally written in Persian. We provide a very short biography of these Muslim scholars, as well as a summary of their main works.

It should be noted that the Arabic word for dissection (tashrih) is used when describing the anatomy of the human body, practising the science of dissection, as well as the forensic sense of autopsy (Savage-Smith, 1995). There is seemingly no explicit support for, or opposition to, either human or animal dissection in the Quran (the "holy book"), Hadith ("the sayings and traditions attributed to the Prophet Muhammad"), and Sunnah (the "customary practices of the very early Muslim community") (Savage-Smith, 1995). The literature largely reports that human dissection was forbidden by the law of Islam, and it is not easy to answer unequivocally whether human anatomical dissection was often practised in medieval Islamic societies or not (Savage-Smith, 1995). However, most Muslim scholars seemingly considered the study of anatomy as a way to increase "their belief in God and to appreciate his wisdom" (Abdel-Halim and Abdel-Maguid, 2003). Moreover, our research and personal observations indicate that in some medical schools in predominantly Muslim countries, the dissection of human bodies is undertaken today; although the majority of bodies used are non-Muslims. As will be seen below, some of the cases where Muslim scholars made anatomical discoveries—relative to Galen's descriptions of 'human anatomy'— suggest that they undertook direct observations of human internal anatomical structures. However, it is difficult to discern, in view of the information available, if those observations were made from systematic dissections, or from practices such as the examination of sick, injured or dead people.

Al-Razi, Abu-Bakr Muhammad ibn Zakariya (Rhazes)
865–925, Al-Rayy, Iran (Persian)

His early interests were in music and art. He then studied alchemy, philosophy, mathematics, and astronomy, as well as medicine. He probably also studied philosophy with his teacher, Ali ibn Sahl Rabban Al-Tabari, a physician and philosopher. He quickly exceeded his teacher and became a famous physician (Souayah and Greenstein, 2005; Amr and Tbakhi, 2007), and was appointed as director of the hospital of his hometown and the capital of the Abbasids (Baghdad).

Died in Al-Rayy on October 27, 925 at the age of 60 (Amr and Tbakhi, 2007; Golzari et al., 2013), being considered one of the two best physicians in medieval medicine (the other being Ibn Sina: see below; Souayah and Greenstein, 2005). He was known among his contemporaries as the 'Galen of his time', and wrote more than 224 books, many of which had a major impact on European medicine (Ahmed, unknown date).

Among his books are *Al-Hawi fi Al-Tibb* (*The Comprehensive Book in Medicine*) and *Ketab Al-Mansuri* (Persaud, 1984; Muazzam and Muazzam, 1989; Souayah and Greenstein, 2005; Shoja and Tubbs, 2007; Golzari et al., 2013). Another famous book is *Al-Shokook Ala Galinoos* (*The Doubt about Galen*), in which Al-Razi criticised some of Galen's theories, especially the four separate 'humours' (essential substances making up the body, roughly explained as blood, phlegm, yellow bile and dark bile), whose balance was thought to be the key to health and a natural body temperature (Amr and Tbakhi, 2007). His clinical observations did not match with Galen's descriptions (Amr and Tbakhi, 2007). He wrote 'What we are reading in Greek books differs in this point from what we see by anatomy' (Najjar, 2010, p1).

Other known books are *Kitab Man la Yahduruhu Al-Tabib* (*Book of Who is not Professional in Medicine or A Medical Advisor for the General Public*), *Kitab Bur' Al-Sa'ah* (*Cure in an Hour*), *Kitab Al-Tibb Al-Rawhani* (*Book of Spiritual Medicine*), *Kitab Al-Judari wa Al-Hasbah* (*The Book of Smallpox and Measles*), *Kitab Al-Murshid* (*The*

Guide), *Al-Syrah Al-Falsafiah* (*The Philosophical Approach*), and *Kitab Sirr Al-Asrar* (*Book of Secret of Secrets*).

He authored the first monograph written on pediatrics, which is known in Latin as *Practica Puerorum* (Amr and Tbakhi, 2007). Importantly, he emphasized the centrality of dissection and of the knowledge of anatomy in the medical field (Abdel-Halim and Abdel-Maguid, 2003). Indeed, he did not allow students to pass a test if they did not answer the anatomy questions correctly, even if they did pass the practical examination (Shoja and Tubbs, 2007; Tubbs et al., 2007).

Two of Al-Razi's (**Fig. 3.3**) most well-known books are: *Al-Hawi fi Al-Tibb* (translated as *The Comprehensive Book on Medicine*) and *Ketab Al-Mansuri* (Persaud, 1984; Muazzam and Muazzam, 1989; Souayah and Greenstein, 2005; Shoja and Tubbs, 2007; Golzari et al., 2013). *Al-Hawi fi Al-Tibb* is known in Europe as *Liber Continens*. It has twenty-four volumes and includes a collection of what Al-Razi learned from Greek and Roman medicine, his own clinical observations and case studies, and techniques of treatment during his years of medical practice. It has been translated into many languages, including Latin (Amr and Tbakhi, 2007). *Ketab Al-Mansuri*, also known as *Liber Al-Mansuri*, is not as extensive and contains extracts from hundreds of previous scholars (Persaud, 1984).

The chapter on anatomy in *Ketab Al-Mansuri* has twenty-six parts, which are divided into sections about structures, such as bones, nerves, muscles, veins and arteries, and organs, such as the eyes, nose, heart, and intestines. In Section 1, Al-Razi gives a general introduction to the human body and describes tendons as a mixture of nerves and ligaments, and the brain as the source of sensory and motor impulses. The section on the skeletal system opens with the skull, which is divided into twenty-three bones (without the teeth), being made up of: six bones in the cranium, fourteen bones in the upper jaw, two bones in the lower jaw, and one other bone called *Al-Watad*. The vertebral column was divided into seven cervical vertebrae and seventeen back vertebrae, including twelve thoracic and five lumbar vertebrae. The sacrum is described as consisting of three fused bones, as is the coccyx, with the last coccygeal element being cartilaginous. He also described the pelvic bone and its attachment to the sacrum, as well as the acetabulum, clavicle, scapula, rib cage, and the upper and lower extremities. He counted four metacarpals and fifteen phalanges, considering the first metacarpal (the thumb) as a proximal phalanx. This is a very interesting point as modern-day authors continue to discuss whether the so-called 'metacarpal 1' is actually a proximal thumb phalanx (see e.g., Reno *et al.*, 2013). Al-Razi further described certain small bones—the sesamoid bones—that fill the gaps in some joints.

Muscles and nerves are covered in Sections 3 and 4, in which it is noted that Galen counted five hundred and twenty-nine muscles (see **Table 2.1**). He described the origin of the spinal cord, its terminus, and the two layers covering it, which extend from the brain's membranes. He mentioned seven cranial nerves and thirty-one spinal nerves, and stated that the nerves had both motor and sensory functions, and that they originate from the brain or spinal cord in pairs. He mainly followed Galen's descriptions of the cranial and spinal cord nerves (Souayah and Greenstein, 2005; Shoja and Tubbs, 2007; Amr and Tbakhi, 2007; Tubbs et al., 2007; Najjar, 2010), although in some respects, his account of the cranial nerves is more similar to modern knowledge (**Fig. 2.9**; Souayah and Greenstein, 2005; Tubbs *et al.*, 2007). Importantly, Al-Razi was the first to describe the recurrent laryngeal nerve as a mixed sensory and motor nerve (**Fig. 2.9**) (Shoja and Tubbs, 2007; Tubbs *et al.*, 2007). Veins and arteries (Sections 5 and 6) are described in detail and Al-Razi followed Galen in considering the liver as the origin of the veins, which he claimed nourish the organs of the body. He outlined the course and branches of the superior and inferior vena cava, and named the major veins in the body, such as the external and internal carotid veins (*Al-Wedaj Al-Haer* and *Al-Wedaj Al-Ghaer*), the cephalic vein (*Al-Ketfi*), and the basilic vein (*Al-Ebti*).

Furthermore, he described details of the circle of Willis, stating: "These two parts (right and left common carotid artery) are divided into two parts (internal and external carotid arteries); one of these parts (internal carotid artery) passes next to the internal jugular vein and ascends to the skull... when it enters the skull, it divides in a wonderful way, and forms something like a network (circle of Willis), which is extended into the brain" (Yarmohammadi *et al.*, 2013a). He stated that arteries originate from the left ventricle of the heart, and described the aorta, coronary arteries, pulmonary veins, and the rest of the arterial system. In Sections 7 and 8, Al-Razi describes the brain and eye, agreeing with Galen's descriptions of the brain having four ventricles (two anterior, one middle, and one posterior). Al-Razi described two nipple-like processes—that is a projection from a structure—in the olfactory bulb. Originating in the anterior ventricle and ending at the ethmoid bone (which resembles a filter), these are responsible for smell. He also described the dura mater and pia mater, and opposed the concept of Galen's that the brain, spinal cord, and ventricles comprise of a single structure (Najjar, 2010).

Within the detailed description of the eye, he included the sclera, choroid, retina, vitreous body, lens, aqueous humors, iris, pupil, cornea, and conjunctiva.

There are brief descriptions of the nose and ear, followed by those of the tongue and salivary glands, pharynx and larynx, chest and lung, and heart in Sections 9–14. Al-Razi stated that the diaphragm divided the region between the clavicles and the pubic bone into two parts: the chest and the abdomen, and also described the mediastinum, the semicircular cartilaginous rings of the trachea, and the mechanism of respiration. He outlined the shape and location of the heart and declared that the aorta and pulmonary veins open into the left ventricle, while the inferior vena cava and the pulmonary artery open into the right ventricle. The heart's valves were also described. Some researchers have reported that Al-Razi opposed Galen's assumptions regarding the presence of a bone at the base of the heart (e.g., Shoja and Tubbs, 2007). However, we found that Al-Razi followed Galen regarding the presence of a bone at the base of the heart in big animals; in small animals it is more like cartilage. He wrote: 'Near its (the heart's) root and origin something looks like the cartilage, which corresponds to a base for the whole heart.' In addition, Al-Razi, like Galen, thought that there was a passage between the left and right ventricles of the heart (Prioreschi, 2006).

Al-Razi continued with the description of the esophagus and the stomach, small and large intestines, liver, and spleen in Sections 15–18. He corrected Galen's description of the stomach having two layers and stated that this organ actually has three layers: longitudinal, circular, and oblique. He further characterised the intestine and the voluntary muscle at the end of the rectum (the anal sphincter), but erred in his description of the liver's vessels, thereby reflecting Galen's theory. He wrote that the portal duct that originates from its concave surface (i.e., the hepatic portal vein) looks like a vessel but does not carry blood, that it has many openings into the stomach and intestine, and brings the food to the liver (through the superior mesenteric vein, inferior mesenteric vein, etc.). He surmised that food becomes blood in the liver, which then moves to the body parts through the great blood vessel that is attached to its convex surface (the inferior vena cava). He considered the splenic and short gastric veins as two tubes connecting the spleen to the liver and the stomach, respectively. In Section 19, the gallbladder, cystic duct, two hepatic ducts, and common bile duct—with its opening into the duodenum—are briefly described, followed by the kidneys and urinary bladder (with two layers) in Sections 20 and 21. While Galen suggested that the ureters run obliquely in the bladder wall and open into its cavity after forming a lid that covers its opening, Al-Razi stated that ureters penetrate the outer layer of the bladder and descend between the two layers, before penetrating the inner layer and entering the urinary bladder. Notably, Al-Razi was one of the few early scholars who described the 'ureterovesical junction' and considered it a great 'urinary anti-reflex mechanism' with the two layers of the bladder adhering to each other to become, as it were, a single layer and squeezing the ureters in between them to prevent urine retention when the bladder is full (Abdel-Halim and Abdel-Maguid, 2003).

He also described the functions of the previously mentioned organs (Section 22), the peritoneum as being supported by eight abdominal muscles, and the omentum, which lies under the peritoneum (Section 23). The penis was characterised as a nervous body, with a large number of cavities inside it and many arteries under it (Section 24). Two channels descending from the peritoneum are reported, which expand to form the internal layer of both testicles (tunica vaginalis). He stated that venous branches go into the testicles and form many convolutions that have a white glandular flesh, which converts the blood in them so that it becomes white. This white blood is transformed to become the 'perfect sperm'. There are two sperm channels leading to the penis (known as the vas deferens). He described the mammary gland as having nerves, veins and arteries, and said that in between them there was white glandular flesh which converted the blood to produce milk (Section 25). The uterus is described with respect to its location and size in different females, and with regards to the three ligaments that hold it in place (Section 26). Like Galen, Al-Razi stated that the uterus has two cavities and a single end. He also mentioned two extensions (the horns of the uterus), behind which are the female testicles (ovaries). These are smaller and flatter than those found in males (Prioreschi, 2006). He wrongly considered the female uterus as analogous to the male penis.

In summary, the literature review on Al-Razi shows that he followed Galen in many of his anatomical descriptions, but that he also opposed, as well as improved upon, Galen's descriptions in many others. For example, he was the first to depict the recurrent laryngeal nerve as a mixed sensory and motor nerve, precisely described the circle of Willis, and distinguished nerves from tendons. He opposed Galen's concept that the brain, spinal cord, and ventricles comprise of a single structure. He also corrected Galen by arguing that the stomach has three, not two, layers, that the coccyx includes three, rather than five bones (three of four is the number commonly given in current textbooks), and established the correct relationship between the ureters and the urinary bladder. These examples of more accurate descriptions by Al-Razi may be a result of detailed observation of surgeries or, perhaps more likely, of human dissections, taking into account their level of detail, and also the fact that he publicly stressed the importance of dissections in the medical field (see, e.g., Al-Razi, ca. 1400–1500, 1674).

Al-Akhawayni Bukhari, Abu Bakr Rabi ibn Ahmad (Joveini)
?–983, Bukhara, Republic of Uzbekistan (Persian)

He was medically trained under one of Al-Razi's students, Abu Al-Qasem Al-Maqanei Al-Razi. He was familiar with the medical knowledge of Hippocrates, Aristotle, Galen, and Al-Razi. He publicly criticised them and opposed some of their theories (Shoja and Tubbs, 2007; Yarmohammadi et al., 2013b).

Unlike the tradition of the time of dedicating books to powerful rulers or wealthy patrons, he dedicated his book to his son and other medical students, and wrote it in an easy and simple style to help beginners (Ardalan et al., 2007; Yarmohammadi et al., 2013a).

The only surviving medical encyclopaedia of *Al-Akhawayni Bukhari*, entitled *Hidayat Al-Mutallemin fi Al-Tibb* (*A Scholar's Guide to Medicine*), is considered the most important Persian text written in the 10th century and the first medical text written in the new Persian language (Farsi Dari) (Dalfardi and Yarmohammadi, 2014). Yarmohammadi *et al.* (2013a, b) stated that this book consisted of three parts and two hundred chapters (*Bab*); while Ardalan *et al.* (2007) wrote that it was arranged in five major sections and one hundred and eighty-four chapters (*Bab*), twenty-eight of which deal with anatomy. The book starts with a theoretical discussion of the elements and humors, followed by several short sections on anatomy and physiology (Ardalan *et al.*, 2007). The cardiovascular, respiratory, and nervous systems, eyes, urinary tract and kidneys, gastrointestinal tract, and other structures are described in detail (Yarmohammadi *et al.*, 2013a, b).

His book, entitled, *Kitab Al-Tashrih* (*Book of Anatomy*), focuses on the anatomical configuration of various body parts and refers the reader to *Hidayat* for further details (Yarmohammadi *et al.*, 2013a). Although Arab rulers at the time did not pay attention to *Hidayat*, Shoja and Tubbs (2007) considered this book a masterpiece of anatomy and medicine for its time.

In *Hidayat*, Al-Akhawayni described the eye and nervous system with extraordinary detail (Shoja and Tubbs, 2007) and stated that nerves originate in the brain and spinal cord, classifying them as either sensory or motor (Dalfardi and Yarmohammadi, 2014). The circle of Willis was explained in greater detail than the Greeks had done, and he suggested that the cranial nerve gives rise to the recurrent laryngeal nerve, which provides cardiac innervation (Shoja and Tubbs, 2007). He described different types of muscles and their origins and insertions, and wrote that some muscles, such as those forming the walls of the bladder, do not have tendons. Dalfardi and Yarmohammadi (2014) consider Al-Akhawayni the first to distinguish between nerves and tendons, but we found that Al-Razi had previously made such a distinction (see above). Al-Akhawayni outlined the structure of the larynx, trachea, lungs, and diaphragm, described the heart in detail, and described, with pinpoint accuracy, the two-layered membrane around it. He also differentiated between arteries and veins in terms of the thickness of their walls (Yarmohammadi *et al.*, 2013b). Despite his agreement with Galen and Al-Razi's theories about the presence of pores in the interventricular septum, he considered blood to move from the right side of the heart to the lungs and then back from the lungs to the left side of the heart. This detail is regarded by some modern scholars as an early description of pulmonary circulation (Yarmohammadi *et al.*, 2013b). In *Hidayat*, the description of the kidney is concise and similar to that found in Greek texts in that it follows Galen's erroneous theory about kidney circulation, *i.e.* that blood moves back and forth between the liver and the kidney (Ardalan *et al.*, 2007).

The authors most cited by Al-Akhawayni were Galen, followed by Hippocrates, Al-Razi, and two other scholars (Ardalan *et al.*, 2007). There are similarities between the anatomy sections of Al-Razi and Al-Akhawayni's books, but the latter has more detail (Shoja and Tubbs, 2007). Therefore, Al-Akhawayni, like Al-Razi, corrected much of Galen's anatomy (e.g., regarding the circle of Willis, distinguishing nerves from tendons, and disproving the presence of a lid at the end of the ureters), but also provided new and accurate descriptions that were not included in Al-Razi's works (e.g., details about pulmonary circulation). It is often suggested that Al-Akhawayni autopsied human bodies, due to his accurate and detailed descriptions (Shoja and Tubbs, 2007; Yarmohammadi *et al.*, 2013a).

Ibn Abbas, Abu al-Qasim Ismail ibn Abbad ibn al-Abbas ibn Abbad ibn Ahmad ibn Idris (Haly/Hali Abbas)

930/949–994, Arejan, Iran (Persian)

Although a Muslim, he did not grow up in a Muslim family like other Muslim scholars, but in a Zoroastrian one. He lived in the Golden Age of Islamic civilization between Al-Razi and Ibn Sina, in the same era as Al-Akhawayni (Fig. 1; Shoja and Tubbs, 2007; Golzari et al., 2013; Zargaran et al., 2013; Dalfardi et al., 2014a).

After finishing his school education in Persia, he moved to Shiraz (southwest of Iran) and started his medical education under the guidance of Abu Maher Shirazi (a prominent Persian physician of that era). Then he moved to Baghdad and became the court physician for Adud Al-Dawlah Fana Khusraw (the king 'Amir' of the Buwayhid dynasty in Persia) (Zargaran et al., 2013; Dalfardi et al., 2014a). While he was in Baghdad, he wrote his medical encyclopaedia *Ketab Kamel Al-Sena-a Al-Tebiah* (*The Complete Art of Medicine*), known also as *Al-Ketab Al-Malki* (*The Royal Book*), which he dedicated to Adud Al-Dawlah (Shoja and Tubbs, 2007; Golzari et al., 2013; Zargaran et al., 2013; Dalfardi et al., 2014a).

Ibn Abbas wrote his medical encyclopaedia *Ketab Kamel Al-Sena-a Al-Tebiah* (*The Complete Art of Medicine*), known also as *Al-Ketab Al-Malki* (*The Royal Book*) while he was in Baghdad (Shoja and Tubbs, 2007; Golzari et al., 2013; Zargaran et al., 2013; Dalfardi et al., 2014a). *The Royal Book* was written in Arabic and has a first, theoretical part, and a second, practical part (Golzari et al., 2013; Zargaran et al., 2013; Dalfardi et al., 2014a). Each part has ten chapters (*Maqala*) and each chapter has numerous sections (*Bab*) (Dalfardi et al., 2014a). The book criticised some famous Greek, Christian, and Muslim physicians, such as Hippocrates, Galen, Oribasius, Paul of Aegina, Serapion the Elder, and Al-Razi. In writing *Ketab*, Ibn Abbas hoped to create the most complete encyclopaedic medical knowledge of the era (Golzari et al., 2013; Dalfardi et al., 2014a).

The anatomical section of the book was the main source for the anatomical sciences in the Muslim world between 1070 and 1170 (Nabipour, 2003; Golzari et al., 2013). It was also highly influential outside the Muslim world. For instance, much of it was translated into Latin by Constantinus Africanus (1015–1087). Strikingly, this translated version, named *Liber Pantegni*, did not make reference to Ibn Abbas, and is possibly one of the oldest examples of major scientific plagiarism (Zargaran et al., 2013; Dalfardi et al., 2014a). In 1127, the entire *Royal Book* was translated into Latin by Stephen of Pisa and entitled *Liber Regius* or *Liber Regalis Dispositionis*. This book was then reprinted in the West and was the main medical reference book at that era (Zargaran *et al.*, 2013; Dalfardi *et al.*, 2014a).

The first chapter of *The Royal Book* has twenty-five sections and includes instructions from Hippocrates and other, older physicians. The second and third chapters are about human organs. The second chapter has sixteen sections, starting with a general discussion about organs and their functions. The organs are divided into two categories: 'similar organs' and 'compound organs'. The former includes structures such as bones, cartilage, nerves, beating vessels (arteries), non-beating vessels (veins), fascia, ligaments, fat, hair, muscles, and skin. Compound organs, on the other hand, include the head, hands, legs, liver, and others, each of which incorporates 'similar organs', i.e., nerves, fat, muscles, veins, and arteries. Section 2 is a general description of bones, which Ibn Abbas considered the hardest 'organ' in the body. He divided the bones of the body into six regions: head, vertebral column, chest and ribs, shoulders and clavicle, and the upper and lower extremities.

The third section details the anatomy—including the shape, borders and function—of the head bones, which Ibn Abbas divided into cranial bones, the upper and lower jaws, and teeth. He illustrated and drew simple shapes for these five sutures of the skull (**Fig. 2.3**) (Ibn Abbas, 1437: 83–84):

> The head has 5 sutures (*Dorowz*) that divided the cranium into six bones. Two of them are not real and three are real. One of the real sutures is located in the front of the head where the crown is usually placed, and it is called coronal suture (*Al-Ders Al-Ekleliy*)... the second real structure lies in the middle of the head and runs longitudinally, being called the straight suture and resembling a sagittal (*Al-mostaqeem Al-shabeeh be-a-sahm*)... the third real suture is located in the back of the head and resembles the letter lambda in Latin, being called Lambdoid (*Lami*). The two 'not real' sutures are located in the two sides of the skull above the ears... they are called squamous sutures (*Al- Qeshriah*) and end at the coronal suture, being equi-distant from the sagittal suture.

Ibn Abbas also named six cranial bones: two square-shaped parietal bones (*Al-Yafokh*), two triangular temporal bones (*Al-Janibain*), the frontal bone (*Al-Jabha*) and the occipital bone (*Azm Moa'akerat Al-Ra'as*). The parietal bones are separated by the sagittal suture. He claimed that the temporal bones are divided into three parts: the stone-like petrous part (*Al-Hajari*), which includes the external opening to the ear; a nipple-like process (or projection), known as the mastoid process, which prevents the lower jaw from sliding away during articulation within the cranium; and the squamous part (*Al-Sadg*), which is not as rigid as the other two parts. The frontal bone is separated from the parietal bones by the coronal suture. The occipital bone is located at the back of the skull. Ibn Abbas also described five bones that protrude from the skull: one is a bone located between the cranium and the upper jaw, called the sphenoid bone (*Al-Watad*), two zygomatic processes on the temporal bone, and two temporal processes on the zygomatic bone. The other four are called the pair (*Al-Zawj*). According to Ibn Abbas, the upper jaw has fourteen bones (seven on each side): six of these relate to the cavities of the eye, two each to the cheeks, the nose, the nostrils, and the upper incisors, respectively (the bones associated with the remaining teeth being located in the cheekbones).

As with Hippocrates, Galen, and Al-Razi, Ibn Abbas stated that the lower jaw in humans (the mandible or *Al-Fak*) is made up of two bones which are joined together in the middle. This is an erroneous statement, and one that is influenced by Galen's descriptions of monkeys. The lower jaw has two projections on each side, one of which is sharp (the coronoid process) and located behind the zygomatic and temporal processes. The other is round and thick (the condylar process) and is located beneath the mastoid process when the mouth is closed. This particular process forms part of the joint and allows for movement of the mandible. He also described the teeth, naming them according to their function, calculating the numbers in adults, and the numbers of roots found in each molar and premolar.

According to Ibn Abbas, the vertebral column is divided into seven cervical vertebrae and seventeen back vertebrae, including twelve thoracic and five lumbar vertebrae. He described the sacrum and coccyx as consisting of three bones each; the last coccygeal bone being cartilaginous, in agreement with Al-Razi. In Section 4, he wrote that the sacrum has two parts:

> The first one is very wide and connected to the last vertebra of the lumbar spine... it has three bones that are similar to the other vertebrae, but two of them are wider, connected to the hip joints, and have two openings in the middle where the nerves exit (not on the sides like the other vertebrae because of the presence of the hip joints)... the second part of the sacral spine is made up from three bones that look like the cartilage (coccyx)... at the end of the third

bone there is an opening that gives out a single nerve (Ibn Abbas, 1437: 88).

He illustrated, in detail, each vertebra in the four regions, including spinous and transverse processes and the configuration of each part, the superior and inferior articular facets, and discussed the importance of each shape.

Section 5 is very short and describes the anatomy of the ribs and how they are attached to the thoracic spine posteriorly, and to the sternum anteriorly, the true and false ribs, and the anatomy of the sternum, which Ibn Abbas depicts as being made from seven cartilaginous bones connected to each other (in this instance, he was, perhaps influenced by Galen's incorrect description of 'human anatomy'). This is followed by a description of the shoulder and clavicle (Section 6), including details on the shape and orientation of the scapula (which possesses three processes, namely the spine, acromion and coracoid, as well as the glenoid cavity, which allows the shoulder to move relative to the body), and the functions of each part. The upper limb (described in Section 7) is divided into three parts: the arm (*Al-Adoud*), forearm (*Al-Saeed*), and hand (*Al-Yad*). Ibn Abbas described the humerus and its anterior and posterior surfaces, the head, medial and lateral epicondyles, trochlea, coronoid fossa, and olecranon fossa. The shoulder joint is discussed in detail—specifically its wide range of movements and its vulnerability to dislocation—as are the forms and functions of the bones making up the forearm, i.e. the radius (*Al-Zend Al-A'ala*) and ulna (*Al-Zend Al-Asfal*), including the supination and pronation of the hands. The twenty-seven bones of the hand are depicted as consisting of a group of eight small bones (the carpal bones) which, together with the radius and ulna, form the wrist joint, and four metacarpal bones that connect the carpal bones to the four fingers, which possess three bones (phalanges) each. Ibn Abbas agrees with Galen and Al-Razi in stating that the thumb is missing a metacarpal and, as with the other digits, has three phalanges. This is a controversial issue that continues to be currently debated. He also discussed the sesamoid bones of the hand and their function.

The lower limbs, Ibn Abbas says, are divided into four parts (Section 8): the thigh, the leg, the foot, and the pelvis, which was considered to be a connection between the lower limbs and the spine. The latter was said to consist of the ilium (*Al-Werk*), ischium (*Al-Kaserah*), and pubis (*Al-Anah*). The foot is depicted as having six tarsal bones, amongst which are the calcaneus (*Al-Aqub*), talus (*Al-Ka'ab*), navicular (*Al-Zawraqi*), as well as four cuneiform bones (wedge-shaped) which, together with a cuboid-shaped bone, are described under a single term (*Al-Rasg*). In addition, Ibn Abbas also identified five metatarsals (*Al-Amshat*), and five toes (*Salamiat*), each toe having three phalanges, except the big toe that has two. This contrasts with his description of the thumb, which has three phalanges. He further described the location and relationship of each of these bones to each other, the ligaments of the foot, and the movement produced by each joint, and finishes this section by summarising the skeletal system of the human body, with reference to its two hundred and forty-eight bones.

Section 9 focuses on the anatomy of the cartilage, which, according to Ibn Abbas, resembles foetal bones. He described ten positions where cartilage occurs, and the role of cartilage in the human body. In Sections 10–13, he identified the function of sensory and voluntary motor nerves in relation to all the body parts except bones, cartilage, ligaments, glands, and fat. He made these exemptions, stating that these tissues/organs do not sense nor are they involved in the movement of other tissues. The nerves originate either from the brain and terminate near organs, or from the spinal cord, ending at organs that are situated some distance from the brain. He also stated that sensory nerves originate from the anterior portion of the brain, while motor nerves originate from the posterior region.

Ibn Abbas (1437: 93–95) described seven paired cranial nerves (**Fig. 2.9**):

> The first cranial nerve travels to the eyes and gives them the sense of sight. The second cranial nerve goes to the eyes and provides motor movement to their muscles. Part of the third cranial nerve goes to the tongue and delivers taste, the other part goes to the temporalis muscle, masseter, the tip of the nose, and the lips, while the last part goes to the gum and teeth with sensation. The fourth cranial nerve divides at the top of the palate and provides taste...
>
> The fifth cranial nerve divides into two pairs when it originates from the brain, the first pair going to the ear's opening and covering it (from the ear drum), providing hearing; the other pair exits from the opening in the petrous bone and joins the third cranial nerve, then they give many branches and innervate the muscles that move the cheeks without the lower jaw and the temporalis muscle. The sixth cranial nerve divides into three pairs: the first goes to the trachea and the base of the tongue and helps the seventh nerve to move the tongue, the second pair goes to the muscle that is located in the

shoulder, and the third pair descends to the neck and the viscera next to the carotid artery (*Al-Sobat*). The seventh cranial nerve goes to the tongue and the muscles of the trachea and provides their movements.

Ibn Abbas did not name the cranial nerves. However, he mentioned the origin of each nerve, the exit point of some of them from the cranium, their final destination, and function. These descriptions have enabled us to identify the aforementioned cranial nerves (**Fig. 2.9**). For example, the first part of the third nerve is the abducens nerve, and the remaining part together with the fourth nerve represents the trigeminal nerve (**Fig. 2.9**). Importantly, it has been reported that Ibn Abbas followed Galen's description of the cranial nerves (Zargaran et al., 2013), but our own direct analysis and translation of Ibn Abbas' original texts indicates that he was the first scholar describing the trochlear nerve (**Fig. 2.9**). Moreover, Ibn Abbas was one of few early scholars who corrected the idea that a single nerve innervates the ear and face, describing instead two separate nerves (vestibulocochlearis and facialis) (Shehata, 2003). In Ibn Abbas' terminology, the sixth nerve corresponds to a combination of the vagus, accessory, and glossopharyngeal nerves. In addition, he named the recurrent laryngeal nerve when he described the third part (*i.e.*, the vagus nerve) of this sixth nerve (**Fig. 2.9**).

He also described the spinal cord and the paired thirty-one spinal nerves: eight cervical, twelve thoracic, five lumbar, three sacral, three coccygeal pairs, and one additional, single nerve at the end of the spine. He mentioned pulsating blood vessels and called them arteries, whose walls are made up of two layers. The outer layer is thin and has longitudinal fibres (tunica adventitia), the inner layer is five times thicker and has circular fibres and some oblique fibres (tunica media). Large arteries have, additionally, an inner layer that resembles a spider's web (tunica intima). As with Galen and Al-Razi, Ibn Abbas also described the veins and erroneously stated that they originate in the liver. He also illustrated the body's arteries, including the aorta, coronary arteries, and the circle of Willis. Ibn Abbas wrote: 'When this vessel [aorta] originates from the heart, two arteries divide off from it; one [right coronary artery] is smaller and moves toward the right ventricle and supplies [blood to] it; the other [left coronary artery] is larger and moves proportionally to the curvature of [the left side of] the heart wall and supplies [blood to] it' (Dalfardi *et al.*, 2014a, p38).

The flesh and fat are described in Section 14, followed by the anatomy of the skin and the membranes in Section 15. Ibn Abbas divided flesh into three types: the first is mixed with nerves and tendons and is considered a compound organ, which is the skeletal muscle. The second type is single flesh and found in three places: the back of the thighs, anteriorly and posteriorly to the spinal column, and between the teeth. The third type (glandular) is related to the glands. He defined a membrane as a thin but hard cover that surrounds the organs. Some organs have a single layer that is attached to them, such as the muscles, while others have double layers, such as the internal organs. He described, in detail, the pleura and pericardium in the thoracic cavity, parietal (*Al-Sefaq*) and omentum (*Al-Tharb*) in the abdominal cavity, and the dura, arachnoid, and pia matter in the cranial cavity. He considered the dura and pia maters as meninges and called them the mothers, but he considered the arachnoid a parietal membrane that protects the brain from the hard dura mater and holds the blood vessels in their places. He reported on the skin (Section 16) and its thickness, the amount of hair on its surface, and its attachment to underlying structures in different areas. Hair and nails are covered in Section 17, which discusses the importance of, and the differences between, the hair, eyelashes, eyebrows, and beard.

His Chapter 3 has thirty-seven sections about compound organs. The first section (Section 1) is a general discussion about these organs, which are divided into three categories: external whole organs, such as the head and limbs; partial compound organs, such as muscles; and internal organs. The second section (Section 2) is about muscles. He identified the muscle as a mixture of red flesh, ligament, and nerve which is covered by membrane. He stated that ligaments originate from bone and are mixed with flesh and nerve, upon which they become part of the muscles. He also described the tendon as a mixture of ligament and nerve without flesh, and explained the mechanism of muscle contraction and relaxation. He stated that muscles have a variety of shapes, sizes, positions, compositions, and a number of tendons attached to them. This is followed by a discussion of the anatomy of the muscular system (Sections 3–9) in a systematic and very detailed manner, starting with the muscles of the head and neck, and followed by the muscles of the hip, lower limb, and foot. He divided the muscles of the face into seven muscles, describing the origin, insertion, and function of some of them. Following his descriptions, we tried to identify the current muscle terminology in **Table 2.2**. All details about his muscle descriptions are given in that table, so here we will refer only to some major points that are important for the context of this paper. His descriptions of the head and neck muscles were not very clear. Moreover, it is apparent that Ibn Abbas mainly followed Galen's muscle descriptions, which, as explained above, were mainly based on the dissections of Old World monkeys (**Table 2.1**). So, for instance, instead of describing a single flexor digitorum longus of the lower limb going to digits 2–5, he described, as Galen did, two muscles, one going to digits 2–3 and the other to digits 4–5; as seen in many monkeys. Moreover, Ibn Abbas' terminology sometimes leads to confusion, particularly concerning the pelvic muscles (see **Table 2.2**). In total, Ibn Abbas reported five hundred and fifty-four muscles in the body.

Table 2.2: Muscular system according to Ibn Abbas. Muscles in the modern terminology ('Modern name') that are marked with question mark (?) indicate that an unambiguous identification of the muscle described by Ibn Abbas was not possible. A plus (+) indicates that the description includes two or more muscles that are close together and might have been mistaken as one muscle. Origin, insertion and function are based on Ibn Abbas books—for clarification modern terms were included in brackets, where we felt it is necessary/helpful. Empty cells indicate unmentioned information (*Al-Ketab Al-Malki* or *The Royal Book*, Ibn Abbas, 1437).

Facial muscles

Ibn Abbas descriptions:	Modern name	Origin	Insertion	Functions
Two wide muscles "*Al-Aredatan*" each one has 4 parts	Platysma + Buccinatorius ?	Spinous process of cervical vertebrae	Edge of the cheek	Movement of cheek, and ear in some people
	Platysma + Risorius ?	From acromion process of scapula and passes by the neck	Ipsilateral edge of lips	Deviation of lips to same side
	Platysma + Depressor anguli oris	Clavicle and ascend	Ipsilateral edge of lips	Pulls mouth obliquely to ipsilateral side
	Platysma + Orbicularis oris	Clavicle and sternum	Contralateral edge of lips "they are decussating"	Narrowing and pursing lips as in blowing
Two muscles that elevate the upper lip	Levator labii superioris			Elevate upper lip
Two muscles that expand the nose	Nasalis + Levator labii superioris alaeque nasi, or Procerus ?			Expand nose
One muscle beneath the forehead skin	Frontalis + Orbicularis oculi			Tightly close and open eyes

Eye muscles (3 groups)

1) Three muscles that move the eyelid	Levator palpebrae superioris	Attached to bone that enclose the eye	Inserted by long tendon that pass in middle of eyelid to the middle edge of eyelid	Opens eyelid
	Orbicularis oculi (orbital and palpebral portions?)	Two muscles originate from two sides of the eye	By two tendons in the two ends of the eyelid	Closes eyelid
2) Support the nerve; people argued with its real number	?			To protect and support soft optic nerve

Ibn Abbas descriptions:	Modern name	Origin	Insertion	Functions
3) Six muscles that move the eyeball	Superior and inferior oblique			Rotate eye
	Superior and inferior rectus			Move eye upward and downward respectively
	Medial and lateral rectus			Move eye to right and left

Muscles of the lower jaw (4 pairs)

Al-Sadgain	Temporalis			Elevates lower jaw
Muscle located inside the mouth	Medial pterygoid (+ Lateral pterygoid?)			Elevates lower jaw
One pair	Digastricus anterior and digastricus posterior	Originated behind ears and moves downward to neck then ascend to chin	Chin	Opens lower jaw
Al-Madegatan, located on the check	Masseter (+ Lateral pterygoid?)			Deviates checks to right and left

Muscles that move only the head

2 pair that bend the head forward	Cleidomastoid and sternomastoid heads of sternocleidomastoideus	Behind ears	Clavicle and sternum	Head flexion
2 pair that bend the head backward	Rectus capitis posterior minor & major	Cervical spine above the joint (NB, he does not specify the joint)		Head extension
2 pair for side bending	Obliquus capitis inferior ? + superior ?	Head joint, on right and left sides		Head side bending

Muscles that move the head and neck

4 pairs located in the back of the head	Longissimus capitis, Semispinalis capitis, Splenius capitis, Splenius cervicis			Bend head and neck backward
1 pair located behind the esophagus	Longus colli	1st and 2nd cervical vertebrae.		Bend head and neck forward and bend head to the side

Muscles that move the throat (pharynx)

2 muscles	Sternohyoideus	Internal side of sternum	Hyoid bone	Pull hyoid bone downward
2 muscles	Sternothyroideus	Internal side of sternum	Thyroid cartilage	Pull cartilage downward

Muscles that move the larynx

2 muscles	Thyrohyoideus ?	From bone that resembles the shield° (thyroid cartilage?)		
4 muscles that attached together	?			Join edges of thyroid cartilage
4 muscles attached to the "unnamed cartilage" (cricoid cartilages)	Cricothyroideus (oblique and transverse?) + Inferior constrictor?			
2 muscles behind the unnamed cartilage	Stylopharyngeus	"Arrow like process" styloid process		Pull arytenoid cartilages backward

Ibn Abbas descriptions:	Modern name	Origin	Insertion	Functions
2 muscles	Arytenoideus			Join arytenoid cartilages together

Muscles of the tongue

2 muscles	Styloglossus	Styloid process	Sides of tongue	
5 muscles	2 Hyoglossus, 2 Genioglossus ?, Geniohyoideus (seen as unpaired muscle?)	Hyoid bone		4 of them move tongue; 5th holds hyoid bone in its place
2 muscles located underneath the tongue, and its fibers are transverse.	Mylohyoideus			

Muscles that move throat

Two muscles one on the right and one on the left	Palatopharyngeus?			Needed to help with swallowing and voice

Muscles that move only the neck

2 muscles on the right side; one anterior and one posterior	Scalene muscles (anterior and/or middle, posterior)			Bend neck to right and front or back
2 muscles on the left side; one anterior and the other posterior	Scalene muscles (anterior and/or middle, posterior)			Bent neck to left and front or back

Shoulder muscles

2 muscles	Trapezius (upper portion)	Vertebrae	Spine of scapula and clavicle	Raise scapula
	Trapezius (middle and lower portion)		Descending and attached to scapula	Protrude scapula
1 thin muscle	Levator scapulae	Transverse process of 1st vertebra	To head of spine of scapula	
1 muscle	Omohyoideus	Hyoid bone	Upper border of scapula at beginning of coracoid process	Tilt scapula towards head
2 muscles	Rhomboideus major and Rhomboideus minor	Thoracic Spinous process		
1 muscle	Serratus anterior?	Humerus	Ascended to shoulder joint until it meets lower parts of lower border of scapula	Move scapula downward and forward; move the arm backward and downward

Ibn Abbas descriptions:	Modern name	Origin	Insertion	Functions

Muscles that move the arm

Ibn Abbas descriptions:	Modern name	Origin	Insertion	Functions
3 muscles	Pectoralis major	Chest, underneath/deep to breast		Move arm toward body
		Upper portion of sternum		
		Sternum		
2 muscles	Latissimus dorsi	Back lower ribs	Inserted by long tendon into arm	
	Latissimus dorsi	Waist		
5 muscles originated from scapula	Infraspinatus	One muscle: from side of scapula		
	Supraspinatus & Subscapularis?	Two muscles: from upper border of scapula		
	Deltoid (middle and posterior portion)	Two muscles		Move arm away from body and backward
1 muscle that occupies the space of the shoulder flesh	Deltoid (anterior portion)?	Clavicles		
1 deep small muscle	Teres minor? Part of Deltoideus or Supraspinatus?			Raise arm with slight rotation

The muscles that move the forearm

Ibn Abbas descriptions:	Modern name	Origin	Insertion	Functions
Two muscles on the anterior side of the arm, the second one is smaller than the first one	Biceps brachii	One originated from deep structures of scapular muscles		Flex forearm
	Brachioradialis	Apparent side of arm from posterior surface	Crossed first one (see above) and inserted on radius	
Two muscle on the back of the arm, their tendons attached to the tendons of the previous forearm flexors[7]	Triceps? Or Brachioradialis?	Medial side of arm	Radius	Extend forearm
	Anconeus?	Arm and extend backward	Ulna	
1 muscle	Extensor carpi radialis longus (+ brevis?)	In middle of lateral epicondyle		
3 muscles	Extensor digitorum, Extensor digiti minimi, Extensor carpi ulnaris	Next to previous one		
3 muscles	Abductor pollicis longus, Extensor pollicis longus, Extensor indicis ?	On side of the 3 previous muscles		
1 muscle	Supinator	Lateral epicondyle	Radius	
2 muscles	Pronator teres, Pronator quadratus			Pronate forearm

Muscles that move the hand and the fingers: a) located on the medial side of the forearm

Ibn Abbas descriptions:	Modern name	Origin	Insertion	Functions
2 muscles in the middle of the forearm, one above the other	Flexor digitorum profundus and superficialis			Flex fingers
1 small muscle above the previous two muscles	Palmaris longus	Medial epicondyle	Its tendon broaden under skin of palm	Protection and support
2 muscles on both side of the previous 3 muscles	Flexor carpi radialis, Flexor carpi ulnaris			
2 muscles below the previous 5 muscles	Pronator quadratus? Supinator? (see above)			Supinate forearm

Muscles that move the hand and the fingers: b) located on the hand

Ibn Abbas descriptions:	Modern name	Origin	Insertion	Functions
Two rows; the upper row has 7 muscles: 5 muscles	4 Lumbricales + Opponens pollicis ?		By small tendons in the first joint after the metacarpal (proximal phalanges)	Tilt fingers upward
1 muscle	Abductor pollicis brevis			Move thumb away from fingers
1 muscle	Abductor digiti minimi			Move little finger away from other fingers
The lower row has 11 muscles; some of them have a common function between the wrist and metacarpal, while others have specific function		Wrist		Common function is: to move the hand
3 of the 11 muscles move the thumb	Opponens pollicis?		One attached to first joint (i.e. joint between metacarpal I & proximal phalanx)	Flex thumb
	Flexor pollicis brevis (2 heads?) + ?		Two attached to second joint (i.e. joint between proximal & distal phalanx)	Move its phalange
Each finger receives 2 muscles from the 11 (the total is 8)	Adductor pollicis? + (3) Palmar interossei & (4) Dorsal interossei (or 8 flexores brevis profundi, if influenced by monkey descriptions of Galen? but then Ibn Abbas should have also counted the 4 intermetacarpales, in a total of 12, not 8, muscles going to the fingers)		First joint (see above)	

Muscles that move the chest

Diaphragm	Diaphragm			Expand chest
2 muscles under the clavicle	Subclavius	End of the clavicle near the scapula	1st rib	Elevate ribs
1 pair	Serratus posterior superior ? (or Scalenus longus ?)	2nd (cervical) vertebra	5th and 6th ribs	Expand chest
1 pair	Serratus posterior superior	Concave side of the scapula	To the back ribs	
1 pair	Serratus posterior superior	7th cervical vertebra		
2 muscles located at the root of the ribs	Iliocostalis			Contract chest
3 muscles	Levatores costarum (longus & brevis), Transversus abdominis?			Pull three floating ribs superiorly
2 muscles extended on the chest on both sides of the xiphoid process	Transversus thoracis			Contract chest
The muscles that lie between the ribs expand and contract the chest	External, inner and innermost intercostales	Those lie between the great part		Expand chest by external fibers and contract it by internal fibers
		Those lie between the cartilage		Contract chest by external and expand by internal

Abdominal muscles

Ibn Abbas descriptions:	Modern name	Origin	Insertion	Functions
8 muscles:				
2 thin muscles above the other muscles, and attached to the skin	Rectus abdominis	Two sides of xiphoid process and lower edge of lower back ribs	Pubic bone	
4 of the 8 are oblique under the first two, two in the right side and two in the left side, they cross each other	External oblique, Internal oblique	Waist bone	Lower ribs	
2 muscles under the previous 4; transverse fibers located above the peritoneum	Transversus abdominis			
Others:				
4 muscle that attached to the testes, 2 on each side "right and left"	Cremaster, Dartos ?			Carry testes
2 muscles attached to the ovaries, one on each side	? (there is only a ligament to ovary, Cremaster poorly developed)			To carry ovaries
1 muscle is a transverse muscle around the neck of the urinary bladder	Sphincter urethrae muscle			Squeeze neck of bladder, help evacuation of urine, and prevent uncontrolled urination.
4 muscles that move the penis	Bulbospongiosus			Elongate and widen the pathway inside penis, erection without deviation
	Ischiocavernosus	Pubic bone	To penis	Erection without deviation, movement to sides
4 muscle in the pelvic floor:	Anal sphincter	The tip of the rectum	Attached to skin	Squeeze rectum
	Pelvic diaphragm	Above the first one and around the tip of the rectum		Hold rectum in its place, narrow it
	2 Levator ani	Twisted, on both sides of second one		Raise pelvis and support rectum

Muscles that move the thigh

Ibn Abbas descriptions:	Modern name	Origin	Insertion	Functions
10 muscles move the thigh, some are located on the iliac bone and others on the ischium, their tendons attached to the hip joint				
2 of the 10 muscles	Iliopsoas?	Has two heads and originate from ilium		Hip flexion, and side tilting
	?	From Ischium		
2 of the 10 muscles	Obturator internus and Obturator externus	Medial side of pubic bone	Both turn around thigh and attached onto each other, then attached to depression near the greater process of femur near knee (medial epicondyle)[8]	Extend thigh and turn it anteriorly
		Lateral side of pubic bone		Extend thigh and turn it posteriorly

MUSLIM SCHOLARS IN ANATOMICAL AND MEDICAL SCIENCES 69

Ibn Abbas descriptions:	Modern name	Origin	Insertion	Functions
6 muscles extend the thigh	Gluteus maximus, Semitendinosus, Semimembranosus, Biceps femoris (long head?), Adductor magnus (hamstring head?) (+Biceps femoris, short head? or one or more of the of the small gluteal muscles, i.e. Piriformis, Gemellus superior, Gemellus inferior, and/or Quadratus femoris? or did Ibn Abbas counted two separate semimembranosus muscles, based on the monkey descriptions of Galen?)	NA	NA	Extend thigh

Muscles that move the leg and foot

3 big muscles on the anterior medial surface of the thigh	Vastus intermedius + medialis	It is a double muscle and could be two muscles; 1) originate from greater protuberance (greater trochanter) 2) anterior surface of thigh	Patella	Extend knee
	Vastus lateralis	Greater protuberance (greater trochanter)	To patella by big tendon then to bone of leg	
	Rectus femoris[a] (and/or Tensor fasciae latae?)	Anterior spine of ilium, straight edge of ilium		
5 muscles behind the medial side	Long head of Biceps femoris? (see above)	Ischium	Lateral side of leg	Move leg to side
	Gracilis	Meeting point of two pubic bone	Medial side of leg	
	Adductor magnus? short head of Biceps femoris?	3 muscles between the first two posteriorly, originate from base of femur	Knee joint by a common tendon	Move knee in different directions
1 small muscle deep in the knee	Popliteus			Flex knee

Muscles that move the foot and toes

7 muscles on the posterior side: 2 of them	Gastrocnemius, two heads	Femur epicondyles	By long tendon in calcaneus	Pull calcaneus, fixe foot, connect leg with calcaneus.
1 muscle	Soleus	Head of fibula	Calcaneus without tendon	Assists the first two muscles
1 muscle	part of Flexor digitorum longus (clearly influenced by monkey descriptions of Galen)	Head of fibula	Its tendon split into two part; goes to 2nd and 3rd toes	Flex 2nd and 3rd toes
1 muscle	part of Flexor digitorum longus (clearly influenced by monkey descriptions of Galen)	Posterior side of the leg	By two tendons pass next to first one, and goes to 4th and 5th toes	Flex 4th and 5th toes
1 muscle	Tibialis posterior	Head of tibia	Its tendon inserted to head from the posterior surface, next to big toe	Flex sole of foot and internally rotate it

Ibn Abbas descriptions:	Modern name	Origin	Insertion	Functions
1 muscle	Plantaris	Lateral epicondyle	By tendon onto calcaneus	
7 muscles on the anterior side: 1 of them	Tibialis anterior	Inner side of tibia	By a tendon into parts proximal to big toe (cuneiform & metatarsal I)	Flex foot and take it off the floor (propulsion in gait cycle)
1 muscle	Extensor hallucis longus	As first one (see above)	Next to first tendon that attached to first bone of big toe	Extend big toe
1 muscle	Extensor hallucis longus?	Between tibia and fibula	Big toe	
1 muscle	Extensor digitorum longus	Inner side of fibula	Four tendon attached to 4 toes	Extend 4 toes
1 muscle	Fibularis longus	Fibula		Flex big toe
1 small muscle that has a small tendon	Fibularis tertius	As previous one		Abduct 5th toe
1 muscle	Fibularis brevis	Fibula	Attached to parts above little toe	Extend foot

Muscles that move the foot

5 muscles in the dorsum of the foot	Extensor digitorum brevis (4) + Extensor hallucis brevis		Attached to 5 toes by five tendons	Move toes to side
7 muscles in the sole of the foot, on the metatarsal bones	4 Dorsal interossei + 3 Plantar interossei			5 muscles move toes to lateral side; 6th and 7th muscles adduct 1st and 5th toes
4 muscles on the tarsal bones	Flexor digitorum brevis			Flex four toes
10 muscles, 2 in front of the first joint of each toe	Flexores breves profundi ? (seemingly influenced by monkey descriptions of Galen)			Flex toe when two muscles contract at same time; Tilt toe when one muscle contracts

He described the brain and divided it into two parts: the cerebrum, which controls sensory and motor functions, and the cerebellum, which is related to movement. The two parts are separated by a thick membrane that penetrates them, forming a fold (tentorium). He described four ventricles of the brain and the corpus callosum, pineal gland, as well as the two meninges that cover the brain: dura matter (*Al-um Al-Jafiah*) and pia matter (*Al-um Al-Raqiqa*), and the spinal cord. Each eye (Section 13) is composed of ten parts; three humours and seven layers. The three humours from outside to inside are the aqueous humour, glacial (lens), and vitreous humour. The seven layers are divided into three groups: (1) retina, choroid, and sclera lie behind the vitreous humour; (2) cornea, pupil, and conjunctiva lie anterior to the aqueous humour; and (3) the arachnoid layer lies between the aqueous humour and the lens. He considered the nose as a pathway for air, but stated that smelling takes place at the end of this pathway, behind the strainer or filter bone (cribriform bone), where the two processes that resemble the nipples are located (the olfactory bulbs). In addition, he stated that there is a connection between the nose and the mouth. The hearing system has three devices: the opening in the petrous bone (external auditory meatus), the membrane that covers the opening (ear drum), and the ear. The membrane is the primary hearing device, which is made up of cranial nerve fibres. The opening has, internally, a spiral shape. The cartilaginous pinna protects the ear and amplifies the voice. The tongue is covered by the same layer that covers the mouth, palate, oesophagus, trachea, and larynx. There are salivary glands and a duct underneath the tongue, and three laryngeal cartilages: thyroid, cricoid, and arytenoid. However, he considered the epiglottis as a mixture of fat, membrane, and glands, forming the main device for the production of sound by closing the larynx. He also described the trachea with its C-shaped cartilage.

The lungs have a separate section that includes their location, number of lobes, thetubes that penetrate them, the membrane around them, and their function: 'these tubes are three: one starts from the right cavity of the heart, other from the left cavity, and the other from the trachea. The one originating from the right cavity is a non-pulsating blood vessel that looks like an artery... it has two thick layers, and it is called the arterial vein (pulmonary artery).' He added that 'the second originating from the left cavity is a pulsating blood vessel but looks like a vein; it has one thin layer and is called the venous artery (pulmonary vein). The third tube originates in the trachea and is made of semi-circular shaped pieces of cartilage, like the trachea... Each of the three tubes divides into four parts when they enter the lungs, two in each lung, but there is a third branch in the right lung. Then they give more branches inside the lungs' (Ibn Abbas, 1437: 130–131).

Galen erroneously described the tracheal semi-circular cartilage as 'rough arteries' and named each of them the *arteria tracheia*, which conveys air (*pneuma*) to the smooth arteries (pulmonary veins) from the heart, and he reported the presence of a pulse in them. Ibn Abbas' *Royal Book* has several pages about the cardiovascular system, including a very detailed anatomical description of the heart and the differences between the arteries and veins (Persaud, 1984). The heart is described as having two main cavities, right and left, unlike Aristotle and Galen, who described three cavities (Leroi, 2014). Like Galen, Ibn Abbas considered the left side of the heart as the origin of arteries, and the liver as the origin of veins, as explained above. He also noted the presence of two atria and two auricles, and described the mechanism of the tricuspid, mitral, and aortic valves, and differentiated between the latter two (Dalfardi *et al.*, 2014a). He further described the pericardium and its relation to the pleura. He mentioned the presence of bone under the base of the heart, like Galen, but was more accurate in reporting that the heart has one apex, which point to the left, while Galen stated that the heart has two apexes, one from each ventricle. Ibn Abbas was the first to pinpoint the presence of a connection between the arterial and venous systems, i.e., the capillaries (Dalfardi et al., 2014a; Daneshfard et al., 2014a). However, he followed Galen's theory about the presence of an opening between the right and left cavities.

Ibn Abbas also described the origin and insertion of the diaphragm, its muscular and tendon parts, its functions, and the oesophageal and caval hiatus. He then moved to the digestive system and described the mouth and its function; the oesophagus and stomach; their boundaries, the oesophageal sphincter, how the oesophagus courses along the thoracic vertebrae, the location of the stomach in relation to other organs, and two layers that make up their walls. However, he stated that the inner layer is nervous in nature and longitudinal in direction, while the outer one is muscular and transverse, similar to Galen's description. He also described the large and small intestine, and the omentum (*Al-Tharb*) as two thick layers derived from the peritoneum and containing blood vessels and fat. He reported on the liver (Section 28) and its location, as well as its relationship to the diaphragm, stomach, and intestine, and stated that the two major veins that originate from the liver are the superior vena cava (*Al-Ajwaf*) and the portal vein (*Al-Bab*). Ibn Abbas stated that the liver has three lobes, unlike Galen and Hippocrates who described five lobes (four lobes are normally described in current textbooks: e.g. Persaud, 1984). Ibn Abbas listed two tubes being attached to the spleen. One receives the atrabilious humour (black bile: the normal atrabilious humour was a sediment of normal blood) and forms in the liver and part of it moves to the blood, while the other part goes to the spleen from the liver (through the splenic vein) (Bekhtiar, 1999). The second tube sends the atrabilious humour to the stomach (through the gastric artery) to stimulate the appetite. The gallbladder has two ducts: one starts from the concave surface of the liver (common hepatic duct) and carries the bile out from the liver's blood, while the other is divided into a bigger part, which is attached to the intestine (common bile duct) and a smaller part, which is attached to the stomach (pancreatic duct).

Lastly, Ibn Abbas described the kidneys and their location, the renal arteries and veins, and the ureter. He followed the mistaken description of Galen instead of the one by Al-Razi in the explanation of the ureters' course through the bladder wall. Ibn Abbas stated that the wall of the urinary bladder has one thick layer and had a round muscle around its neck. When the ureters penetrate the wall of the urinary bladder, they form a lid-like membrane that covers the end of each ureter and moves in one direction. This allows the urine to exit the ureter without the possibility of retention. In males, there is an extra process in the neck of the urinary bladder that ends in the penis. *The Royal Book* has a detailed description of the uterus, its location, and its size during different situations, its relation to the rectum and urinary bladder, and its function. Ibn Abbas stated that the wall of the uterus has a single layer that is composed of many fibres running in different directions—longitudinal, transverse, and oblique—and has two horns (fallopian tubes) that are connected to two female eggs (ovaries) that are smaller and harder than the male eggs (testes). Similar to Galen and Al-Razi, Ibn Abbas stated that the uterus has two cavities that share a single vagina, and which are important when women have twins. He also stated that male foetuses occupy the right portion of the uterus while female foetuses occupy the left side. He wrote that the female breast consists of flesh and glands that are nourished by two arteries and two veins divided from the pulsating vessels and the vena cava when they reach the clavicles. The male testes consist of flesh and glands, and are covered by a membrane derived from the peritoneum. Each one receives a vein from the liver and arteries from the great blood vessel, which are divided into small branches. The male seminal vesicles are longer, wider, and harder than those in females, and the penis is a long, hollow organ originating from the pubic bone and has two muscles on its sides. He stated that males and females have the same reproductive organs, which simply differ in their configuration and temperatures.

In summary, it can be said that the two chapters on human anatomy in Ibn Abbas' *Royal Book* include more extensive descriptions than those in Al-Razi's book (*Al-Mansuri*). As with Al-Razi, Ibn Abbas' work contains statements which concur with some of Galen's inaccurate descriptions and others which correct and improve upon them. Some of these statements are seemingly influenced by earlier Muslim scholars, including Al-Razi himself (e.g. that the human coccyx does not include five bones). Other accounts concern new data that was not provided by authors such as Al-Razi, including the description of a trochlear nerve, of two separate nerves for the face and ear (facialis and vestibulocochlearis), and of the capillaries connecting the arterial and venous systems. The description of this new data may be based on direct observation from the dissecting of human cadavers by Ibn Abbas himself, although there is no solid historical evidence that he systematically performed such practices, at least in public.

Ibn Sina, Abu Ali Husain ibn Abdullah (Avicenna)
980–1037, Afshaneh, near Bukhara (Persian)

He was a great physician and teacher, and one of the most important scholars of the ancient world (Naderi et al., 2003; Afshar, 2010; Najjar, 2010). In the West he was known as 'the prince of physicians' (Afshar, 2010), 'Aristotle of Islam' and the 'Second Doctor' (after Aristotle, 'the First') (Golzari et al., 2013).

When he was five, he moved with his family to Bukhara and by the age of eight he memorised the Quran in full. It is said that he was skilled in many sciences, including geometry and mathematics, by the age of ten, and that as a teenager he was encouraged to study medicine by Abu Sahl Al-Masihi (Golzari et al., 2013).

He started his professional career as a physician when he was seventeen and cured the ruler of Bukhhara of an unknown disease that other physicians could not treat. As a result he was rewarded with access to the royal library of Bukhara (Shoja and Tubbs, 2007; Golzari et al., 2013). In 999 he moved to Gorganch of the Khwarizmi Kingdom, now known as Urgench in Uzbekistan, where he became familiar with Al-Biruni (973–1048), a great pharmacist and polymath. He travelled to various Persian cities and had access to many libraries. When he was in Hamadan, he was *wazir* (minister) of the ruler Shams Al-Dawla until 1021. After the death of the ruler, he was detained by the new ruler, Samaa Al-Dawla, on suspicion of disloyalty.

While in prison, he wrote the medical treatise *Kitab Al-Qawlanj* (*The Treatise on Colic*). After Alaa Al-Dawla Ibn Kakuya's conquest of Hamadan, he was released from prison and went to Isfahan (city in central Iran) where he wrote *Kitab Al-Shifa* (*Book of Healing*). On the way back to Hamadan in 1037, he died at the age of fifty-eight from a severe colic, perhaps due to stomach cancer.

He wrote his famous medical encyclopaedia *Al-Qanun fi Al-Tibb* (*The Canon of Medicine*) in Jurjan (a city north east of Tehran) and finished it in Ray, in 1020 (Shoja and Tubbs, 2007; Golzari et al., 2013). This book was highly influential in the history of medicine and was translated into Latin, Hebrew, German, French, and English. It was the core of European medical science between the 13th and 18th centuries because of its superiority to Galen's medical book (Naderi et al., 2003; Afshar, 2010; Gadelrab, 2010; Najjar, 2010; Golzari et al., 2013; Dalfardi et al., 2014b). To this day, the *Canon* remains the most impressive written book in medicine, according to many scholars (Afshar, 2010). Ibn Sina carefully read all the previous major anatomical works and added his own observations (Naderi et al., 2003), many of which continue to be accepted today (Persaud, 1984; Naderi et al., 2003).

Ibn Sina (or Avicenna: **Fig. 1.2**) wrote around two-hundred and seventy different publications, many of them about medicine (Nabipour, 2003). He wrote his famous medical encyclopaedia, *Al-Qanun fi Al-Tibb* (*The Canon* *of Medicine*) in Jurjan (a city north east of Tehran) and finished it in Ray, in ~1020 (Shoja and Tubbs, 2007; Golzari et al., 2013). The *Canon* was divided into five parts: Part 1) chapters about general anatomy and principles of medicine; Part 2) dedicated to *materia medica*; Part 3) diseases of the special organs; Part 4) general medical conditions; and Part 5) formulary (Golzari et al., 2013).

The first chapter was about the skeleton, had thirty sections, and focused on bones and joints, and their functions. Sections 2 and 3 dealt with the cranium and skull bones, in which he described, as with previous scholars Al-Razi and Ibn Abbas, three 'real' and two 'unreal' sutures (**Fig. 2.3**). He divided the bones of the head into five parts. Four can be considered as the walls (frontal, two parietal and temporal bones, and one occipital bone), with the fifth, the sphenoid bone, as the base. His descriptions of the bones of the lower and upper jaws, and the nose (Section 4) were similar to those of previous scholars, including the erroneous statement that adult humans have two, separated, mandibles. The teeth were described in Section 5 as the only sensitive bones, following Galen: 'experience proved that teeth have sensitivity, they have the power that comes from the brain to distinguish hot and cold.' In Sections 6 and 7, Ibn Sina explained the roles and functions of the vertebral column, and described in detail the vertebral, intervertebral and transverse foramina, superior and inferior articular processes, spinous and transverse processes, and superior and transverse costal facets. He illustrated and described seven cervical, twelve thoracic, and five lumbar vertebrae, and the sacrum and coccyx with three vertebrae each, following the explanations of Al-Razi and Ibn Abbas (Sections 8–12). In Section 13, he provided a general discussion about the vertebral column and its importance. Sections 14 and 15 contained detailed descriptions of the ribs and sternum, as well as the clavicle, its function, location, and connections with the sternum and shoulder (Sections 16 and 17). He illustrated the scapula, acromion and coracoid processes, the spine of the scapula, glenoid cavity, and acromioclavicular ligament. He then described the humerus (Section 18), and the shoulder joint as a loose joint covered by four ligaments: the capsular ligament, subscapular tendon, and the biceps brachii tendons. He illustrated the bicipital groove, medial and lateral epicondyles, olecranon fossa, and trochlea, described the forearm, and described and illustrated the radius and ulna (Section 19). He stated that the radius is smaller and responsible for pronation and supination, while the ulna is larger and responsible for flexion and extension.

The structure and mechanisms of movement of the wrist joint were part of Section 20, followed by detailed illustrations of the carpal, metacarpal bones, and phalanges (Sections 21–23). He divided the carpal bones into two rows, like Ibn Abbas, but his first row had three bones (scaphoid, lunate, and triquetrum), while the second had four (trapezium, trapezoid, capitate, and hamate). Ibn Sina stated that the eighth bone (pisiform) was created to protect the nerve located next to the palm (palmar portion of the ulnar nerve). Like Al-Razi and Ibn Abbas, Ibn Sina considered metacarpal 1 as a thumb proximal phalanx (Ibn Sina, 1597; Bakhtiar, 1999). Section 24 described the nails, while Section 25 focused on the ilium, ischium, pubic bones, and acetabulum. He mentioned the functions of the lower limbs and illustrated the anatomical features of the femur (head, medial and lateral epicondyle), tibia, and fibula (Sections 26–28). In Section 29, he described the knee joint as an articulation between two processes that lie at the lower end of the femur, and the two concave articular surfaces of the upper end of the tibia. He described the cruciate ligaments, the tibial and fibular collateral ligaments, and the patella, as well as the twenty-six bones of the foot, including five metatarsals and fourteen phalanges; the three cuneiforms and cuboid were described under a single name (*Al-Rasg*). He concluded the skeletal system by counting a total of two hundred and forty-eight bones in the human body, excluding the hyoid and sesamoid bones (Ibn Sina, 1597; Bakhtiar, 1999).

Chapter Two was about the muscular system and contained thrity sections called 'art'; each 'art' having an anatomical description of organs, followed by many sections about the diseases and the treatments of each organ. Section 1 described muscle as a combination of nerves, ligaments, and flesh: when a muscle contracts it pulls the tendon, which is made from nerve and ligament, and bends the body, and when it expands, the tendon is released, so that the part returns to its place (Ibn Sina, 1597; Bakhtiar, 1999). However, Al-Qattan (2005) argued that Ibn Sina opposed Galen and recognised muscles and nerves as different structures. As with Ibn Abbas, we correlated Ibn Sina's muscle descriptions with modern terminology in **Table 2.3**, in order to compare these descriptions with those provided by Galen. As all the details are given in this table, in the text below we refer simply to some points that are crucial in the context of the present paper. Basically, as with Ibn Abbas, Ibn Sina mainly followed Galen's muscle descriptions, thus perpetuating many of Galen's erroneous descriptions of 'human myology' based on monkey dissections. For instance, he described a flexor digitorum profundus with tendons extending to digits 1–5, an extensor indicis extending to digits

2 and 3 and an extensor digiti minimi extending to digits 4 and 5, and two muscles, instead of a single flexor digitorum longus going to digits 2–5 in the lower limb (see above). However, there are some cases in which Ibn Sina descriptions are more accurate than those of Galen, *e.g.*, when he described the sternomastoid and cleidomastoid as heads of the sternocleidomastoideus, and not as separate muscles. However, this muscle is similar in monkeys and humans, so this seems to reflect more a difference in the interpretation of what a muscle is compared to a muscle head, and is, thus, not necessarily an indicator that Ibn Sina dissected sections of the human muscular system (Ibn Sina, 1597; Bakhtiar, 1999).

Table 2.3: Muscular system according to Ibn Sina. Muscles in modern terminology ('Modern name') that are marked with question mark (?) indicate that an unambiguous identification of muscle described by Ibn Sina was not possible. A plus (+) indicates that description includes two or more muscles that are close together and might have been mistaken as one muscle. Origin, insertion and function are based on Ibn Sina book (*Canon of Medicine*)—for clarification modern terms were included in brackets, where we felt it is necessary/helpful. Empty cells indicate unmentioned information (*Al-Qanun fi Al-Tibb* or *The Canon of Medicine*, Ibn Sina, 1597)

Forehead muscle (1 muscle)

Ibn Sina descriptions	Modern name	Origin	Insertion	Functions
Forehead is moved by thin broad membranous muscle that extends under its skin	Frontalis			Raise eyebrows when contracted and help eyes closure when relaxed
Eye muscles (6 muscles)				
Four muscles located on four sides: top, bottom, medial, and lateral sides	Rectus superioris, Rectus inferioris, Rectus medialis, Rectus lateralis			Each muscle moves eye to its direction
Two muscles placed obliquely	Obliquus superior, Obliquus inferior			Rotate eye
Muscle behind eyeball	?			Support optic nerve
Muscles eyelid				
Two muscles in lateral corners of upper eyelid	Orbicularis oculi			Close eyelid
One muscle in middle of upper eyelid	Levator palpebrae superioris			Opens eyelid

Ibn Sina descriptions	Modern name	Origin	Insertion	Functions
Muscles that move cheeks and lip				
One muscle in each side called "Al-Are-datan" (platysma) that has four fibers	1) Platysma + Depressor anguli oris	Clavicle	Ipsilateral edge of lower lip	Pull mouth obliquely to ipsilateral side
	2) Platysma + Orbicularis oris	Clavicle and sternum	Contralateral edge of lips (they are decussating)	Narrowing and pursing lip as in blowing
	3) Platysma + Risorius ?	Acromion process of scapula	Ipsilateral edge of lips	Deviation of lips to same side
	4) Platysma + Buccinatorius ?	Spinous process of cervical vertebrae	Parts of cheek	Move cheek and lips; moves ear in some people
Muscles that move lips (previous muscles + 4 muscles)				
1 pair (2 muscles)	Zygomatic major + minor?	Highest point in check	Edge of lips	When contracted move lips to their side
1 pair (2 muscles) from below	Depressor anguli oris? Risorius?			
Muscles that move ala of nose (2 muscles)				
2 small strong muscles mixed with check muscles	Levator labii superior alaeque nasi	Cheekbone	Nose	Move wings of nose
	Nasalis muscle			
Muscles that move lower jaw				
2 muscles (1 pair) called Al-Sadg; located behind zygomatic arch	Temporalis	Bone above brain (temporal bone)	Lower jaw by long tendon	Close lower jaw
Two muscles inside mouth	Medial pterygoid + Lateral pterygoid?		Descend to lower jaw	Help previous one to close lower jaw
Opening muscles	Digastric	Process that located behind ear and resemble needle (styloid process)	Chin	Open lower jaw
Two triangular muscle, one on each side of check and has two heads	Masseter		One head into lower jaw and other into zygomatic arch	Chewing
Muscles of head (2 groups)				
Muscles that bend head only are two; some may consider each as two muscles because its end has two heads	Sternocleidomastoideus	Behind ears	Clavicle	Flex head to its side when one muscle of one side moves; bend head forward in a moderate way (flexion) when muscles of two sides move together
Muscle that bends head and neck together; one pair located behind esophagus and reaching cervical vertebrae 1 and 2	Longus capitis + Longus colli		1st and 2nd cervical vertebrae	Flex head only when part near esophagus contracts; flex neck when part that attaches to two vertebrae contracts

MUSLIM SCHOLARS IN ANATOMICAL AND MEDICAL SCIENCES

Ibn Sina descriptions	Modern name	Origin	Insertion	Functions
Muscles that extend head only are 4 pairs located behind previous muscles. They all originate above joint (between head and cervical vertebra 1): (1) One pair attaches to transverse process of C1 and lies above another muscle; (2) that attach to spinous process of C2; (3) Another muscle originated from transverse process of C1 and attached to spinous process of C2; (4) 4th pair originate above and descend obliquely below 3rd pair and attach to transverse process of C1	1) Obliquus capitis superior		Transverse process of 1st cervical vertebra	First two pairs bend head backward (extension) without tilting or with slight tilting
	2) Rectus capitis posterior major		Spinous process of 2nd vertebra	
	3) Obliquus capitis inferior	Transverse process of 1st cervical vertebra	Spinous process of 2nd cervical vertebra	Contraction of 3rd or 4th pair bend head to its side while contraction of both bend head without tilting
	4) Rectus capitis anterior or Rectus capitis lateralis	From above	Transverse process of 1st cervical vertebra	Bend head backward with tilting
Muscles that extend head and neck are 3 deep pairs and one superficial pair. Each muscle form one side of a triangle whose base in bone of skull and other two sides descend to neck	Longissimus capitis Semispinalis Splenius capitis	One pair descends along sides of vertebrae		Extend head and neck
		One pair tilts to transverse processes		
		One pair lies centrally between sides of vertebrae and transverse processes		
	?			
Two pairs of muscles that bend head to sides. These four muscles are small but are functional due to "their proper place" and secure position by other muscles above them	Rectus capitis anterior or Rectus capitis lateralis[10]	Anterior joint of head	2nd cervical vertebra	Each one bends head to its side; if two muscles on one side contract together they bend head to one side (right or left) while if two anterior or two posterior sides contract together they help in flexion or extension of head
	Rectus capitis posterior minor?	Posterior joint of head	1st cervical vertebra	

Muscles that move larynx

One of the muscles that open throat	Thyrohyoideus muscle	Hyoid bone	Thyroid cartilage	Pull arytenoid cartilage forward and above, so larynx expands
Pharyngeal muscle that pulls pharynx down and is considered as a common muscle between larynx and pharynx	Sternothyroideus	Internal side of sternum	Thyroid cartilage	
Another pair attach to previous one	Sternohyoideus			
Two pairs, one of them attaches to posterior side of arytenoid cartilage	Posterior Cricoarytenoid + Lateral cricoarytenoid?[11]	Arytenoid cartilage		Raise and pull back arytenoid cartilage so it moves away from thyroid cartilage, allowing larynx to expand
One pair of muscles go to both sides of arytenoid cartilage	Arytenoideus	Sides of arytenoid		Separate arytenoid cartilage from thyroid cartilage, expanding larynx

Ibn Sina descriptions	Modern name	Origin	Insertion	Functions
One of muscles that narrow larynx originates from hyoid bone and attaches to thyroid cartilage then become wide and turns to cartilage that has no name until two muscles unite behind cartilage that has no name	Thyrohyoideus?	Hyoid bone	Cartilage that has no name	Narrows larynx
Four muscles thought to be two double muscles that connect two edges of thyroid cartilage with cartilage that has no name	Straight part and oblique parts of Cricothyroideus ?			Tighten lower part of larynx
Muscles that close larynx and are located inside it	Thyroarytenoideus	Base of thyroid cartilage	Edges of arytenoid cartilage	Pull arytenoid cartilage down and close larynx

Muscles of pharynx (2 pairs)

Ibn Sina descriptions	Modern name	Origin	Insertion	Functions
One pair has been mentioned in larynx section	Sternothyroid	See above	See above	See above
Other pair	Sternohyoideus + Middle pharyngeal constrictor ?	Sternum	Ascend and attach to hyoid then to pharynx	Pull it down

Muscles of throat:

Ibn Sina descriptions	Modern name	Origin	Insertion	Functions
2 muscles located in throat	Stylopharyngeus ? Palatopharyngeus ?			Help in swallowing

Muscles of hyoid bone (3 pairs)

Ibn Sina descriptions	Modern name	Origin	Insertion	Functions
One pair	Mylohyoideus? Genio-glossus?	Sides of lower jaw	Straight line on hyoid bone (body of hyoid bone)	Move hyoid bone towards lower jaw
Another pair originates below chin and passes below tongue	Geniohyoideus?	Under chin	Upper edge of hyoid bone	Draws bone toward sides of lower jaw
	Stylohyoideus	Bone that resemble arrow located near ear (styloid process)	Lower part of straight line on hyoid bone	Raise hyoid

Muscles that move tongue (9 muscles)

Ibn Sina descriptions	Modern name	Origin	Insertion	Functions
Two wide (transverse) muscles (1 pair)	Styloglossus	Bone that resemble arrow (styloid process)	Sides of tongue	
Two tall (longitudinal) muscles (1 pair)	Hyoglossus	Upper part of hyoid bone	Base of tongue	
Two muscles move tongue obliquely	?	"Lower rib" (greater horn) of hyoid bone	Tongue between tall and wide muscles (see above)	Move tongue obliquely
Two muscles reverse tongue and located below previous muscles. Its fibers extend transversely under these muscles and it is attached to entire jaw bone	Mylohyoideus			
Single muscle connect tongue and hyoid bone	?			Takes one to other (tongue and hyoid bone)

MUSLIM SCHOLARS IN ANATOMICAL AND MEDICAL SCIENCES 81

Ibn Sina descriptions	Modern name	Origin	Insertion	Functions
Muscles that move neck only (2 pairs)				
One pair in right side and one pair in left side	Right scalene (anterior, middle, and/or posterior?)			Contraction of one muscle leads to side bending with rotation to that side. Contraction of two muscles of same side lead to side bending only (right or left). Contraction of 4 muscles keeps head straight.
	Left scalene (anterior, middle, and/or posterior?)			
Muscles that move chest				
Diaphragm, located between respiratory organs and digestive organs	Diaphragm			Expand chest and elevate ribs
2 muscles under clavicle	Subclavius	Top of shoulder (acromion)	1st rib in right and left sides	
1 pair, each muscle has two heads: (1) highest one attaches to neck and moves it; (2) lowest one attaches to chest and moves it; mixed with muscle inserted onto ribs 5 and 6, that we will discuss later	?			
One pair located in concave side of scapula that is attached to pair of muscles descending from cervical spine to scapula; they become one muscle when they insert onto false ribs	Serratus anterior and Levator scapulae		Inserted onto false ribs	
One pair	Serratus posterior superior	7th cervical vertebra and 1st and 2nd thoracic vertebrae	True ribs	
One pair extended below upper ribs	Transversus thoracis			Contract chest
On edges of upper ribs and attach to sternum between xiphoid process and clavicle and joins abdominal muscles	Sternalis (variation in humans)			
Two pairs of muscles that are "helpers"	Serratus posterior inferior			
4 muscles between each 2 ribs; outer muscles are expanders while inner muscles are contractors	External, inner, and innermost abdominal muscles			Expand and contract chest
Muscles that move shoulder joint:				
Three muscles come from chest	Pectoralis major, abdominal head	Below/deep to chest	Anterior side of arm (humerus) near edge of clavicle	Move arm toward chest with slight lowering movement (depression?) which followed by scapula
	Pectoralis major, clavicular head	Upper portion of sternum	Surrounds inner side of head of humerus	Move arm toward chest with slight elevation

Ibn Sina descriptions	Modern name	Origin	Insertion	Functions
	Pectoralis major, sternocostal head	Big double muscle originate from entire sternum	Lower portion of anterior surface of humerus	Move arm toward chest with elevation when upper portion contracts and with depression when lower portion contracts
Two muscles: (1) one is greater and originate from ilium bone and false ribs; (2) the other is smaller and originates from skin of ilium, not from bone	Latissimus dorsi	Ilium and false ribs	Inserts deeper than great double muscle that originates from sternum (pectoralis major)	Move arm toward false ribs in straight direction
	?	Skin around ilium	Joins tendon of muscles that originate below/deep to breast	Like previous one but with slight backward tilting

5 muscles originate from scapula:

1) One muscle	Supraspinatus	From scapula and occupy space between spine of scapula and upper edge of scapula	Upper portion of outer side of head of humerus	Move arm away from body with tilting
Two muscles: 2) One sends its fibers to lower side of spine of scapula and occupies space between spine and lower border; 3) Other one attaches to first one as part of it but does not attach to upper portion of scapula	Infraspinatus	Upper border of scapula	Outer side of head of humerus	Move arm away from body with tilting
	Teres minor		Inserts obliquely on outer side of humerus	
4) 4th muscle occupies concave side of scapula	Subscapularis		Its tendon attaches to internal portion of inner side of head of humerus	Backward rotation of humerus
5) Another muscle	Teres major	Lower edge of lower border of scapula	Its tendon inserts above great muscle originate from ilium (latissimus dorsai)	Pull up head of humerus
Humerus has another muscle that has two head and two actions (one for each head) and a common action; one head is internal and other is external	Deltoid	From lower surface of clavicle and from neck (?)	It surrounds head of humerus and inserts near great muscle that originate from chest (pectoralis major)	When two heads contract they raise arm in straight direction

Muscles that move arm

Extensor muscle has two folds	Long and lateral heads of triceps	Lower anterior part of humerus and lower edge of scapula	Inner side of elbow	Extend elbow with inward deviation
	Medial head of triceps	Posterior side of humerus	Outer side of elbow	Extend elbow with outward deviation and extend elbow in straight direction when two heads contract
Flexor muscle is two folds and first is greater; second has two fleshy heads	Biceps	Highest part of rim of glenoid (long head) and coracoid process (short head)	By nervous tendon into front side of radius	Flexion with inward deviation
	Brachialis	One head from anterior side of humerus and other one from posterior side	They form a cover for humerus in their way and they are inserted onto ulna	Flex with outward deviation when two heads contract; flex forearm in a straight direction

MUSLIM SCHOLARS IN ANATOMICAL AND MEDICAL SCIENCES

Ibn Sina descriptions	Modern name	Origin	Insertion	Functions
Supinator muscles of forearm is pair	Supinator	One of them is located outside between two bones of forearm (radius and ulna)	Radius without tendon	Supinate forearm
	Brachioradialis	Other is thin and long, originated from lateral epicondyle of humerus	By a membranous tendon in inner side of radius	
Pronator is pair located outside	Pronator teres	Medial epicondyle of humerus	Radius	
	Pronator quadratus	Shorter and transverse, originates from ulna	End of radius near wrist joint	

Muscles that move wrist

Ibn Sina descriptions	Modern name	Origin	Insertion	Functions
Among extensor muscles there is a muscle attached to another muscle as if they were a single muscle.	Extensor pollicis brevis ?	Middle portion of ulna	Its tendon attach to thumb	Abduct thumb
Other muscle	Abductor pollicis longus ?	Radius	First carpal bone near thumb (trapezium)	Supinate wrist
Another muscle extend over outer side of radius	Flexor carpi radialis? Or extensor carpi radialis (with wrong insertion stated on columns to the left)	Epicondyle of humerus (medial epicondyle)	By two-heads tendon into middle of metacarpal bone, in front of middle and index fingers	Expand and supinate wrist
Extensor muscle is pair located on outer side of forearm: 1) Lower muscle; 2) Upper muscle	Flexor carpi ulnaris	Medial epicondyle of humerus	Metacarpal of little finger	Supinate hand
		Higher than previous muscle	Same insertion	
Another muscle	Flexor carpi radialis?	Lower portion of humerus between previous two muscles	It has two ends that cross each other and insert between index and middle fingers	When these muscles act together they bend wrist

Muscles that move fingers

Ibn Sina descriptions	Modern name	Origin	Insertion	Functions
Extensor muscle located in middle portion of outer side of forearm	Extensor digitorum	Prominent portion of humerus epicondyle	By 4 tendon into 4 fingers	Extend 4 fingers
Three muscles join together and located next to previous muscle	Extensor digit minimi (based on wrong monkey description of Galen)	One originate from middle potion of lateral epicondyle	By two tendons into little and ring fingers	Tilt fingers downward
	Extensor indicis (based on wrong monkey description of Galen)	One double muscle originate from lower epicondyles of humerus and edge of ulna	By two tendons into middle and index fingers	
	Extensor pollicis longus	Upper portion of radius	Thumb	
Three flexors muscles above each other and located centrally: (1) First one is hidden under others; (2) second one is smaller and above first muscle; (3) third muscle is extended into palm and it is not flexor	1) Flexor digitorum profundus (based on wrong monkey description of Galen - monkeys have no flexor pollicis longus)	Upper portion of ulna and radius	Its tendon widens and divided into 5 tendons to inner side (palmar) of 5 fingers	Flex 1st and 3rd phalanges of four fingers and 1st and 2nd phalanges of thumb
	2) Flexor digitorum superficialis	Medial epicondyle of humerus	Middle joints of 4 fingers	Flex 4 fingers
	3) Palmaris longus			Give hand sensitivity, prevent hair growth, and support and strengthen palmar surface of hand

Ibn Sina descriptions	Modern name	Origin	Insertion	Functions
Muscles that located in hand are 18 muscles in two rows				
7 muscles in lower row; 5 of them here	4 Lumbricales and Abductor pollicis brevis	One belongs to thumb, originates from first of carpal bones (scaphoid)		Bend fingers upward
6th is short and wide and has oblique fibers	Adductor pollicis	Metacarpal of middle finger	Thumb	Tilt thumb downward
7th is located near little finger	Abductor digiti minimi + Flexor digiti minimi brevis?	Metacarpal of little finger		Tilt little finger downward
11 muscles in lower row located below extended muscle that was known to Galen (palmar aponeurosis)				
8 muscle are	Flexores breves profundi (based on monkey descriptions of Galen)		Each two attach to 1st joint of 4 fingers, one above other	Flex first joint
3 muscles belong to thumb	Flexor pollicis brevis (2 heads?) + Opponens pollicis ? + ?			One flexes first joint and two flex second joint
Muscles of back				
Muscles that extend spine are two and we believe that each one composes of 23 muscle because each one receives oblique muscle from each vertebra except the 1st vertebra	Semispinalis ?			Moderate contraction will erect spine, while tight contraction will hyperextend spine and when one side contract spine bend toward that side
Flexor muscles are 2 pairs: one superior and moves neck and head and located on sides of esophagus	Longus colli muscle	Head and neck	Upper 5 or 4 thoracic vertebrae	
Other pair located below previous one	Psoas major muscle	From 10th and 11th thoracic vertebrae and descends		
Muscles of abdomen (8)				
One pair descends in straight direction	Rectus abbominis	Xiphoid process	Pubic bone	Evacuate feces and urine from viscera, and in fetus from uterus; support diaphragm during exhalation and during contraction of chest; keep stomach and intestines warm
Two muscles crossing previous one in straight angle and located above membrane which extended over abdomen (peritoneum)	Tranversus abdominis			
Two oblique pairs cross each other; located above transverse pair	External oblique muscle	Spinous process	Pubic bone	
	Internal oblique muscle	Ilium	Xiphoid process	
Muscles of testes				
4 muscles: one pair (2 muscles) for each testes	Cremaster Dartos ?			Protect and carry testes
One pair for female testes (ovaries)	Cremaster ?[12]			

Ibn Sina descriptions	Modern name	Origin	Insertion	Functions
Muscles of urinary bladder				
One muscle around its opening	Sphincter urethral muscle			Prevent involuntary evacuation of urine
Muscles that move penis (2 pairs)				
One extend on sides of penis	Bulbospongiosus			Widen pathway and elongate penis when they extend
Other pair	Ischiocavernosus	Pubic bone	Base of penis in oblique direction	Moderate contraction leads to straight erection while strong contraction leads to strong erection to back; contraction of one muscle leads to deviation to its side
Muscles of pelvic floor (4)				
One muscle attach to anus and to muscles of pelvis	Anal sphincter			Tightens and closes anus
Another muscle located above previous one	Pelvic diaphragm		Root of penis	
One oblique pair above previous muscles	Levator ani muscle			Carry pelvis
Muscles that move thigh:				
Extensor muscles				
One of the greatest muscles in the body, extends thigh and covers pubic and hip bones and surrounds inner and posterior sides of femur until it ends at knee. It has different origins and different actions	Adductor magnus	1) Some fibers originate from lower portion of pubic bone; 2) Other fibers originate higher than previous one; 3) Some fibers originate from ischium	Knee	1) Thigh extension with medial tilting; 2) Thigh Elevation with medial tilting; 3) Thigh extension in straight direction
Another muscle covers hip joint from posterior side and has three heads and two ends; two are fleshy and one is membranous	Gluteus maximus ?	3 heads originate from ilium, ischium, and coccyx	Two ends attach to posterior side of head of femur (greater trochanter)	Contraction of one end leads to hip extension with deviation to its side; contraction of two heads leads to straight extension
Another muscle	Gluteus medius ?	Outer surface of ilium	Top of greater trochanter	Extension with medial rotation
Another muscle like previous one; attaches first to lesser (?) trochanter then descends and does like previous one	Gluteus minimus		It attaches first to lesser trochanter then descends and does like previous one	As previous muscle but is less involved in extension and more in rotation
Another muscle	Superior and inferior gemellus	Originates below ischium bone and tilts backward		Extension with slight backward rotation

Ibn Sina descriptions	Modern name	Origin	Insertion	Functions
Flexor muscles:				
Straight muscle descends from two origins	Iliopsoas	Lumber vertebrae and ilium	Medial small head of femur (lesser trochanter)	Flexion with slight deviation to medial side
Another muscle	Pectineus	Pubic bone	Lower portion of small head of femur (lesser trochanter)	
One extends next to previous muscle in oblique direction as portion of great muscle (Adductor magnus)	Adductor longus			
4th muscle	Rectus femoris	Upright erect process in ilium (anterior inferior iliac spine)		Leg extension and hip flexion
Tilting and rotation muscles (some of them have been mentioned with extensors and flexors)				
Very long muscle	Gracilis or long head of Adductor magnus?	Pubic bone	Reaches knee	Inward tilting
Two muscles externally rotate thigh; one originates from wide bone	Piriformis	From wide bone (sacrum)		Outward tilting
	Gemellus superior/inferior? Qadratus femoris?			
Two rotator muscles	Obturator internus Obturator externus	One from inner side of pubic bone and other from outer side	Hole near greater trochanter	Rotation; contraction of one rotates thigh to its side with slight extension
Muscles that move leg and knee				
Three muscles that move knee joint are located in front of femur and extend knee				
One has two heads and two ends	Vastus lateralis?. Vastus medialis?	One head from greater trochanter	Fleshy end attach to patella then become tendon	Extend knee joint
	Vastus intermedius	Other head from anterior surface of femur	Membranous attach to medial side of femur (?)	
Other has been mentioned with thigh flexors	Rectus femoris	See above	See above	See above
3rd muscle	Vastus lateralis?. Vastus medialis?	Greater trochanter	It joins second muscle (Rectus femoris) and forms one tendon that surround patella then attach to beginning of leg	Extend knee
One extensor	Gracilis	Meeting point of pubic bone (symphysis pubis)	Descend obliquely along medial side of femur and inserts onto tibia	Extend leg with medial tilting
Another muscle mentioned in some dissection book that located in opposite side, in lateral side, and descends obliquely. It is the most oblique muscle in body	Tensor fasciae latae ?	Ischium bone	It reaches non-fleshy location (?)	Extend leg with lateral tilting; when two muscles contract together (this muscle and previous one) they produce straight extension

Ibn Sina descriptions	Modern name	Origin	Insertion	Functions
Flexor muscles				
Long narrow muscle near extensor muscle; originates from process of ilium (rectus femoris) and passes obliquely towards medial edge of tibia	Sartorius	Ilium and pubis (?)	Medial edge of tibia	Extend leg and tilt foot toward groin
Three: medial, lateral, and middle muscles. Lateral and middle do flexion with lateral tilting, and medial muscles do flexion with medial tilting				
One of medial muscles	Semitendinosus	Base of ischium	Passes obliquely behind femur until it reaches medial side of tibia and attach there	Flexion with medial tilting
Other two medial muscles	Biceps femoris (short and long heads)	Base of ischium bone	Lateral side of tibia	
One muscle hidden in knee fold	Popliteus			
Muscles that move ankle joint (dorsiflexors and planterflexors)				
One of dorsiflexors located anterior to tibia	Tibialis anterior	Lateral side of head (?) of tibia	Descends obliquely toward big toe and attaches near base or big toe	Dorsiflexion of foot
Another muscle	Fibularis tertius	Head (?) of lateral bone (fibula)	Near base of little toe (5th toe)	Dorsiflexion of foot
One of planta flexors is pair	Gastrocnemius	Heads of femur (medial and lateral epicondyles)	Descend and fill back of leg with flesh; a great tendon originates from them and attaches to calcaneus	Plantarflexion of feet; fixation of foot on ground
Another muscle that helps previous one	Soleus	Head of fibula	Descends and attaches without tendon to calcaneus above (?) previous muscle	Help previous muscle
Another muscle has two heads	Tibialis posterior + Flexor hallucis longus ?	Head of tibia where it meets fibula (inner side of head)	Descends between two bones (tibia and fibula) and splits into two tendons; one attaches to cuneiform bone near big toe; other tendon passes 1st one and attaches to 1st joint of big toe	One head plantarflexes foot; other extends (?) big toe
Muscles that move foot				
Muscles located on posterior side of leg				
One of flexors of toes	Flexor digitorum longus (two distinct muscles based on Galen's monkey descriptions)	Head of fibula and descends onto its surface	Its tendon divides into two tendons that flexes 3rd and 4th toes	Flex 3rd and 4th toes
Other muscle is smaller		Posterior side of tibia	Its tendon divides into two tendons that flexes 2nd and 5th toes	Flex 2nd and 5th toes
3rd muscle has been mentioned before that originates from head of tibia where it meets fibula (inner side of head)	Tibialis posterior + Flexor hallucis longus	See above		

Ibn Sina descriptions	Modern name	Origin	Insertion	Functions
Muscles located on food				
10 muscles located on foot were discovered by Galen. Each toe receives two muscles: one in right and one in left	Flexores breves profundi (based on Galen's monkey descriptions)			Flex toes when they contract together; tilting to one side when one muscle contracts alone
4 muscles located on tarsal bones: one muscle for each toe	Flexor digitorum brevis			
2 special muscles for little and big toes	Abductor hallucis + Abductor digiti minimi ?			Flexion
5 muscles located on dorsum of foot	Extensor digitorum brevis + Extensor hallucis brevis			Tilt toes to lateral side
5 muscles below previous 5	(4) Lumbricales + Abductor hallucis muscle?			Tilt toes to medial side

His Chapter Three has six sections about the nervous system, including how the cranial (**Fig. 2.9**) and spinal nerves traverse from the neck to the coccyx. Ibn Sina considered the optic nerves to decussate at the optic chiasm, opposing Galen who said that they meet without decussating. Chapter Four contains five sections about the vascular system. First, he described the arteries (Section 1), their two layers, and their origin from the left side of the heart. He characterised the venous arteries (pulmonary veins) as arteries that originate from the thinnest part of the heart, and which branch and distribute into the anterior parts of the lungs (Section 2). Unlike other arteries, pulmonary veins have only one layer. Pulmonary arteries were defined as originating from the posterior part of the lungs. The second artery originating from the left side of the heart is the greatest artery (the aorta); the coronary arteries and their courses being considered branches of the aorta. The ascending aorta (Section 3) divides into a larger branch on the right side (brachiocephalic), which, in turn, divides into three branches. The right and left common carotid arteries ascend with the internal jugular veins; with the right subclavian artery dividing until it reaches the top of the scapula and upper limbs, while the smaller branch (left subclavian artery) extends to the axilla and divides like the third branch (right subclavian artery) of the great branch (brachiocephalic artery). A detailed description of the common carotid artery, including internal and external carotid arteries and the circle of Willis, are given in Section 4, followed by an account of the descending aorta and its branches and the veins, in Section 5. He stated that the veins originate in the liver, like Galen and previous Muslim scholars. He described the portal vein (Sections

1–3) as originating from the concave surface of the liver and delivering nutrition to it, while the other vein that originates in the liver (inferior vena cava) delivers nutrition to the body's organs (Ibn Sina, 1597; Bakhtiar, 1999).

The anatomy of the compound organs was included in Part Three of the *Canon* at the beginning of each section, where the diseases and treatment of each organ are discussed. The brain is the first compound organ (Section 2), its anterior part being seen as softer because it is the origin of the sensory nerves. Its posterior part (cerebellum) was seen as harder because it is the origin of motor nerves. These two parts are separated by a tough membrane (the dura mater) that is gradually inserted between them and forms a fold (tentorium). He described the ventricles as cavities filled with pneuma and mentioned three ventricles (a paired anterior, one middle, and one posterior). He also characterised the olfactory bulb, dura mater, pia mater, pineal gland, venous sinuses, colliculi, pituitary gland, and infundibulum. He described the anatomy of the eye (Section 3) and detailed the names and functions of its parts (sclera, choroid, retina, vitreous body, lens, iris, aqueous humour, cornea, conjunctiva, and pupils). He referred to the ear (Section 4) as the organ for hearing and the auricle as a curved shell that collects sounds, and stated that the ear has a channel in the petrous bone that is spiral in shape (cochlea) to increase the distance when air moves inside it. The inner surface of the ear is covered by the acoustic nerve, which comes from the fifth cranial nerve (facial and vestibulocochlear nerves) (**Table 2.1**).

The nose (Section 5) has two pathways that terminate at a bone resembling a filter (the cribriform plate of the ethmoid bone), which lead to two processes that look like nipples (olfactory bulb). The membrane of the brain (dura mater) is also pierced at this place to allow odours to penetrate through. Interestingly, Ibn Sina described the nasolacrimal duct and reported the presence of two channels between the inner corner of the eyes and the inner side of the nose, which explains the tasting of tears. In Section 6, he characterised the mouth and tongue, and stated that the ventral surface of the tongue is continuous with the oesophagus and inner side of the stomach. The dorsal surface is divided by the median groove into two parts, which run parallel to the sagittal suture, and are attached to each other. The tongue has soft flesh that is penetrated by small vessels and four nerves. Glandular flesh is at the root of the tongue and produces saliva (sublingual gland), and there are two opening where the saliva comes out (the sublingual and submandibular openings).

Lips are described as a mixture of flesh and nerves, and cover the mouth and teeth, retain saliva, help with speech in humans, and are related to beauty.

The pharynx was characterised as a space where there are channels for breath and food, and where the uvula, both tonsils (palatine tonsils), and epiglottis are located (Section 9). The uvula is a hanging fleshy substance on the upper part of the larynx and is important in letting air in gradually, so that it does not suddenly hit the lungs with its coldness. It also prevents the entry of smoke and dust. The uvula is also an instrument used for speech, being like a door that closes the outlet of the voice. Tonsils are fleshy organs, protruding at the root and superior to the tongue; like two little ears. The epiglottis is a membranous piece of flesh attached to the palate and lying close to the trachea. A bone (hyoid bone) was described above the epiglottis with four 'ribs' (horns), two of which are superior and two of which are inferior. The trachea (Section 10) is built up from numerous incomplete cartilaginous rings, and the oesophagus is located posterior to it, at a point where the trachea becomes membranous. The rings are connected by ligaments and covered with a membrane. The two main bronchi and their branches pass through the lungs and accompany the branches of arteries and veins. The larynx is an organ responsible for the completeness of the voice and holding one's breath, while the lungs contain branches of the trachea, the venous artery (pulmonary vein), and the arterial vein (pulmonary artery). These branches are connected by soft, loose and airy flesh, and each lung has two left and three right lobes.

The heart (Section 11) is seen as being made of strong flesh and, as with Galen's descriptions, Ibn Sina wrote that the base of the heart acts like a root and resembles cartilage, thereby, providing it with a solid support. The heart has three chambers, two large, lateral ones and one in the middle, with a channel between them. Arteries arise from the left chamber and have two layers, the inner layer being harder than the outer one. The wall of the left chamber is thicker than the right chamber. The breast (mammary gland) is the part that produces milk and is composed of veins, arteries and nerves (Section 12). The gaps between them are filled with a glandular flesh that has no sensitivity and which is white in colour. The oesophagus (Section 13) is composed of flesh, an inner longitudinal membranous layer and an outer transverse membranous layer that facilitate the downward pushing of food. He described its location, course, and the cranial nerve that descends next to it (vagus nerve). Ibn Sina considered the oesophagus to widen gradually until it reached the stomach, whereupon it became part of it, because the two organs have the same layers, but with the oesophagus being made of a muscular substance while the stomach being more of a nervous substance. Additionally, the inner layer of the stomach has some oblique fibres that are missing in the oesophagus. The duodenum was not considered part of the stomach, as it does not have the same layers. He described the course of a branch of a cranial nerve (vagus) that gives the stomach the hunger sensation, and described the peritoneum and omentum (Section 13).

The liver (Section 14) has red flesh-like blood and resembles coagulated blood. It does not contain nervous fibres, but the veins that originate in it branch inside it like fibres. The liver absorbs chyle from the stomach and intestines via a branch of the portal vein, which he designated as mesenteric veins originating from the concave surface of the liver. The chyle is cooked in the liver and converted into blood. This then spreads through the body via the vena cava, which originates in the convex surface of the liver. The liver also sends water to the kidneys, yellow bile to the gall bladder, and black bile to the spleen. He also described the hepatogastric, hepatoduodenal, and hepatocolic ligaments, and the suspensory ligaments of the liver, as well as the gallbladder and spleen (Section 15), stating that the gallbladder is a sack attached to the liver at the side of the stomach. It has a nervous layer and 'mouth' that opens into the liver and absorbs yellow bile. The gallbladder sends one branch (common bile duct) to the duodenum and sometimes a small branch to the end of the stomach. The spleen was characterised as being an elongated, tongue-shaped organ that is attached to the left posterior side of the stomach. It absorbs black bile from the liver through a tube between them (splenic vein).

Six intestines (Section 16) are described: duodenum (twelve or *Al-ethna Ashar*), jejunum (fasted or *Al-saem*), ileum (small intestine or *Al-Deqaq*), ceacum (*Al-a'awar*), colon (*Al-colon*), and rectum (*Al-mostaqim*). All these parts are attached to the spine by (mesenteric) ligaments. The kidney (Section 18) is the organ that drains water from the blood. The right kidney is higher than the left kidney, and there is a cavity (renal pelvis) inside each kidney, and two ureters connecting the kidneys to the urinary bladder. According to Ibn Sina, the wall of the urinary bladder (Section 19) is nervous, providing strength and elasticity. He described the urethral sphincter and its role, the two layers of the wall of the bladder, and the diagonal course of the ureters between these two layers. He stated that in males the urethra is longer and has three curves, so it does not drain completely, while in females it is straight and short. The testis (Section 20) has a cavity and produces sperm. According to him, in most males the right testicle is stronger than the left one, and he described the inguinal canal and the epididymis (seen as part of the seminal vessels, and being attached to the testis but looking as if it is not).

The penis is described as a combination of nerves, ligaments, vessels, and flesh, originating from the pubic bone. It is full of cavities that are collapsed when they are empty, and has large arteries and many sacral nerves. Lastly, Section 21 described the uterus. The ovaries were described as "female testis", being smaller and more circular than the male ones. Each ovary has its own sack that is nervous. Females also have seminal vessels (oviducts) but these are shorter, originate in the ovaries, extend to the ilium like two horns, and are attached to the uterus. He described a nervous, circular belt inside the uterus, and stated that the uterus is attached to the spine by strong ligaments, and to the umbilicus, bladder, and the wide bone (sacrum) by loose ligaments as well. According to him, some women have a two-cavity uterus while others have a one-cavity uterus. He also described the location of the uterus and its relationship to visceral organs, its two layers, the muscular wall of the vagina, and stated that the uterus is similar to the penis, but is internal.

The sources of Ibn Sina's descriptions are still controversial. Most of his anatomical descriptions are similar to those of Galen and previous Muslims scholars (e.g., the hyoid bone, the muscular system, etc.). Furthermore, he quoted Galen in many sections of his book. However, his descriptions are remarkably detailed and include some new data, which has led some authors to argue that he might have secretly performed human dissections, while others considered that his anatomical considerations might be mainly based on his clinical observations (Shoja and Tubbs, 2007).

Ibn Al-Haytham, Abu Ali Al-Hasan ibn Al- Hasan (Al-Hazen)
965–1040, Basra, Iraq (Arab)

Born in Basra of Old Persia (now southern Iraq) in about 965, he became an Arabic Muslim polymath, physician, astronomer, mathematician, physicist and philosopher. His primary education began in his hometown and continued in Baghdad. He was the central figure in the history of Islamic optics and was known as the 'Father of Optics' (Lindberg, 2003; Unal and Elcioglu, 2009; Daneshfard et al., 2014b).

He was employed as a judge in Basra. Then, several religious movements with diverse and conflicting views appeared, which disillusioned him and led to his decision to spend his time and efforts to study science (Daneshfard et al., 2014b).

Due to his extensive knowledge of mathematics and physics, he became particularly famous in Iraq, Syria and Egypt. As a result, Fatimid Caliph Al-Hakim (the ruler) summoned him to move to Cairo to create an engineering plan to control the Nile river floods. As his plans for flood regulation were unworkable he got arrested and confined to a house for ten years until Al-Hakim's death in 1021. During this time, he wrote several influential texts on various scientific issues (Daneshfard et al., 2014b).

After being released from house arrest, he resided in a building close to Cairo's Azhar Mosque and taught sciences (mainly physics and mathematics), wrote scientific books, and made money by copying texts. He died in Cairo in 1039 or 1040, after writing a total of more than 200 books and novels, of which only 90 are believed to be preserved (Unal and Elcioglu, 2009; Daneshfard et al., 2014b).

He addressed different topics, including Aristotelian natural philosophy, and Galenic anatomy and physiology, but his highest achievements were in mathematics, being considered by some scholars as one of the most important mathematicians in the history of science (Lindberg, 2003).

On the subject of optics, Ibn Al-Haytham wrote, in Arabic, a lengthy book entitled, *Kitab Al-Manazir* (*Book of Optics*), which was influential in the Islamic world and Europe until the 17th century (Lindberg, 2003; Hehmeyer and Khan, 2007; Unal and Elcioglu, 2009). His goal was to take the entire optical tradition, separate the truth from error and blend the truth into a single successful account of the phenomena of light and vision (Lindberg, 2003). *Book of Optics* had the earliest preserved diagrams of the eyes and their connection to the brain (**Fig. 2.4**; Contadini, 2007). It had seven chapters (*Maqalat*), each with several sections (*Bab*). The fifth section of the first chapter contains detailed anatomical descriptions of the eye, including his famous diagram of it (**Fig. 2.4**; Daneshfard et al., 2014b). The diagram shows a horizontal cross section of the eyes, the optic chiasm, and a simple statement at the top saying that the two nerves are connected to the brain. The nose is shown to point out to the location of the eyes (Contadini, 2007). He stated that the eye is made up of layers, membranes, and different objects, and that it originated and started from the forebrain. He described two hollow optic nerves, the optic chiasm, and the two membranes that cover them, which, according to him, are extensions of the brain's membranes (dura mater and pia mater). The optic canal was characterised as being the exit of the optic nerve from the cranium (Sabrah, 1983).

Like previous scholars, he described the sclera, conjunctiva, choroid, retina, vitreous body, lens, iris, pupil, aqueous humours, and cornea, as well as their function and importance, in detail. However, there is a fundamental difference between his descriptions and those of previous scholars: he combined information from physics, mathematics and anatomy. For example, he stated that, although the lens is the middle structure of the eye, it

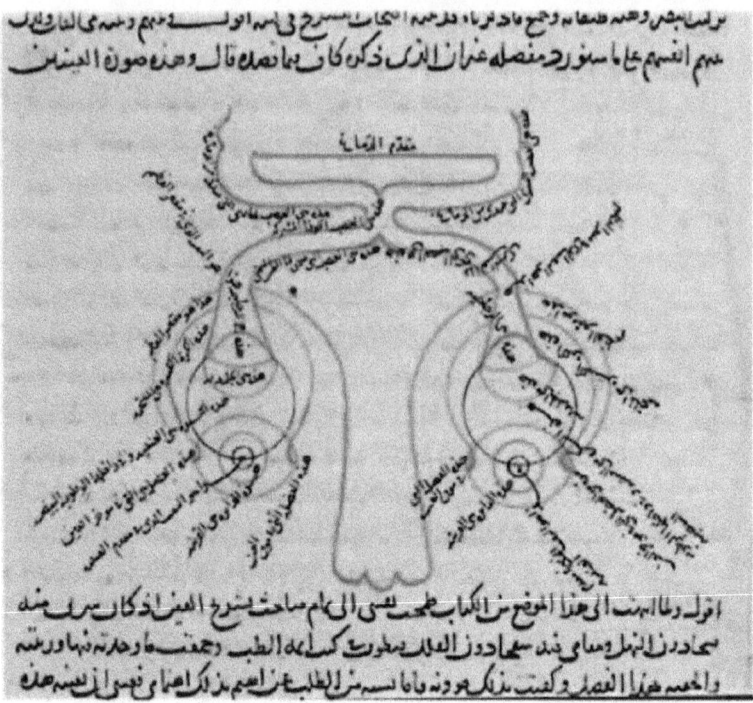

Fig. 2.4. Ibn Al-Haytham's (965–1040) diagram of eye anatomy from his book of optics (Kitab Al-Manazir).

is not always located in the middle, but can shift to the anterior portion of the eye at a specific distance. He also argued that the two sides of the cornea are parallel to each other, indicating that the cornea has the same thickness along its length. He concluded that if a straight line was directed through the centre of the cornea, pupil, and the lens was extended posteriorly, it would pass into the middle of the optic nerve. He also noticed that the parts of the eye move as one piece where the optic nerve exits the cranium. All this data, and more, introduced his new theory of vision, which formed the basis for future knowledge in this subject. Importantly, this theory of vision proposed that light emanated from objects and not the eyes, therefore opposing the ideas of previous scholars such as Galen, Ptolemy, and Euclid (Daneshfard et al., 2014b). He tried to explain how the ocular anatomy could function as an optical system and, probably for the first time in the history of medicine, presented the idea of a magnifying role for the convex ocular lens and of the function that the eyes—and not the brain—have in light perception (Daneshfard et al., 2014b).

In summary, Ibn Al-Haytham's descriptions of the eye show that he was aware of all the relevant knowledge of his time. Importantly, he reported new data about the anatomy and function of the eye that was crucial for the advancement of anatomy and medicine, not only in the Muslim world, but also in the West (Unal and Elcioglu, 2009).

Ibn Rushd, Abu Al-Walid Muhammad ibn Ahmad ibn Muḥammad (Averroes)
1126–1198, Cordoba, Spain (Arab)

His father was a judge (Qadi) and his grandfather was the grand judge of Cordoba. He was known as the grandson (Al-Hafeed) because his grandfather carried the same name (Muhammad).

Known as 'the prince of science', he was a religious scholar and polymath writing about the Quran, Islamic law, philosophy, medicine, astronomy, mathematics, physics, and geography (Savage-Smith, 1995, 2005; Tbakhi and Amr, 2010). He was introduced to the court by Ibn Tufayl, the philosophic minister of the second Almohad Amir or Caliph (prince), and was appointed as a judge in Seville, Spain, in 1169.

In 1171 he moved to Cordoba and was appointed as a judge for ten years (Tbakhi and Amr, 2010). He moved to Marrakesh in 1182, and worked as the personal physician of the Caliph, and then moved back to Cordoba and became the chief judge. The theologians raised an opposition against his writing; as a result, he fell out of favour with the Caliph and was accused of heresy, interrogated, and banned to Lucena (a town in southern Spain).

Most of his books were burned at the Caliph's request, except the ones on medicine, arithmetic, and elementary astronomy. When the banishment was remitted, he was called back to Marrakesh, where he died on December 10, 1198 (Tbakhi and Amr, 2010).

He was a prolific and encyclopaedic author, and wrote more the 20,000 pages (Tbakhi and Amr, 2010). Today, only a few of his abundant writings survive in the Arabic language (Savage-Smith, 2005). He wrote twenty books on medicine, the most important one, written between 1153 and 1169, being called *Al-Kulliyat Fi Al-Tibb* (*Generalities or General Medicine*), known in Latin as *Colliget* (Muazzam and Muazzam, 1989; Savage-Smith, 1995; Tbakhi and Amr, 2010). It was translated into Latin and Hebrew ninety years later and widely used for teaching in the West until the 18th century.

Ibn Rushd, also known as Averroes, wrote twenty books on medicine, the most important one, written between 1153 and 1169, being called *Al-Kulliyat Fi Al-Tibb* (*Generalities or General Medicine*), and known in Latin as *Col-*

liget (Muazzam and Muazzam, 1989; Savage-Smith, 1995; Tbakhi and Amr, 2008). *Generalities* was divided into seven books. The first contained the anatomy of organs (*Tashrih Al-a'ada'a*) and had twenty-five chapters. In Section 1, he divided organs into similar and compound organs, like the previous scholars. He counted and briefly described the bones (Section 2) and stated that the head has fifty-five bones: including six in the cranium, fourteen in the upper jaw, two in the lower jaw, one under the cranium called *Al-Watad* (pterygoid), with sixteen teeth in each jaw. He stated these fifty-five bones are attached to each other by sutures except the two bones in the lower jaw, which are connected by a joint, as was reported by Galen and previous Muslim scholars. He described seven cervical and seventeen back vertebrae (twelve thoracic and five lumbar), three fused sacral vertebrae, and three fused coccygeal vertebrae. He stated, as previous Muslim scholars, that the last coccygeal vertebra is cartilaginous (Besteiro and Morales, 1987).

He reported a connection between the head and the first cervical vertebrae (atlas), and a joint between the sacrum and the hip bones (Section 2). Then he characterised the shape, location, and connections of the clavicles, sternum, scapula, and ribs. He counted thirty bones in the upper extremities, and followed Galen and previous scholars in considering that the first metacarpal bone is the proximal phalange of the thumb. He reported twenty-seven bones in the lower limbs, and concluded this section with a quotation from Galen stating that the human body has two hundred and forty-eight bones plus the sesamoid bones, hyoid bone, and a cartilaginous bone at the base of the heart. Sections 3 and 4 covered the beating and non-beating vessels (arteries and veins, respectively). Similar to Galen and other previous scholars, he stated that the left side of the heart is the origin of arteries, while the liver is the origin of veins. He distinguished between arteries and veins by the number of layers (three layers vs. two layers) and described the course and termination of the blood vessels in the head and neck, chest, both extremities, abdomen, and pelvis.

He described the spinal (peripheral) nerves and their origins, destinations, and branches (Section 5), counting thirty-two pairs and one single spinal nerve at the end. Seven cranial nerves were described (**Fig. 2.9**). He distinguished between the ligaments and tendons by their structure and function, and stated that the ligaments originate from the edges of articulated bones and their nature is an intermediate between bones and nerves. Tendons, on the other hand, originate in nerves that innervate the muscles and are attached to a bone, and their nature is an intermediate between ligaments and nerves. He divided the body's flesh into three types: flesh with nerves and tendons that form muscles; only flesh, which is located in the thigh, spine, and between the teeth; and glandular flesh that fills the spaces in the testis, breast, and base of the tongue. Muscles were the first compound organ described by him (Section 6). He stated, as Galen did, that there are five hundred and twenty-nine muscles in the body, but did not describe the origin and insertion of each of them (Besteiro and Morales, 1987).

He described the brain (Section 7) with regards to its shape, the cavities inside it (four ventricles), the olfactory bulb (consisting of two processes resembling a nipple), the ethmoid bone (the bone that resembles a filter) and the two membranes around it (dura mater and pia mater). The eye (Section 8), he said, has seven layers: sclera, choroid, retina, arachnoid layer, iris, cornea, and conjunctiva; and three humours: vitreous body, lens, and aqueous humour. He named and described the pupil and its role in vision under dark and light conditions. The nose has two pathways at its terminus (Section 9): one opens into the mouth, while the other opens into the bone that resembles a filter (cribriform plate of the ethmoid bone). These pathways are concealed by a thick cover, like the one that lines the mouth. Ears were described briefly as sinuous tubes in the stone-like (petrous) bone (Section 10). At the end of this tube is the fifth cranial nerve (**Fig. 2.9**), which forms a membrane that extends over the petrous bone (eardrum). The tongue was illustrated as a piece of loose, white flesh with many small vessels (Section 11). Underneath it there are two openings originating from glandular flesh. The pharynx and mouth were mentioned together (Section 12), and Ibn Rushd stated that the end of the mouth has two openings: the one in front is the air pathway, pharynx, and trachea, while the one in the back is the food/drink pathway, i.e., the oesophagus. The pharynx has a 'valve' that controls its opening and closure, being a vocalisation machine with three cartilages and various muscles, which were briefly characterised.

Galen (129–200/216) (Souayah and Greenstein, 2005)	Al-Razi (865–925)	Ibn Abbas (930–994)	Ibn Sina (980–1037)	Ibn Rushd (1126–1198)	Vesalius (1514–1564) (Souayah and Greenstein, 2005)	Modern terminology
	(olfactory bulb)	(olfactory bulb)	(olfactory bulb)	(olfactory bulb)		I. Olfactory
I. Optic	I. Optic	I. Optic	I. Optic	I. Optic	I. Optic	II. Optic
II. Oculomotor, VI Abducens	II. Oculomotor, VI Abducens	II. Oculomotor, (Trochlear)[a]	II. Oculomotor, VI Abducens	II. Oculomotor (VI Abducens?)	II. Oculomotor, Trochlear, Abducens	III. Oculomotor
						IV. Trochlear
III. Trigeminal, IV. Trigeminal	III. Trigeminal, IV. Trigeminal	III. Trigeminal: V1, V2, V3[b]; Abducens IV. Trigeminal (V2)	III. Trigeminal, IV. Trigeminal	III. Trigeminal, IV. Trigeminal	III. Trigeminal, IV. Trigeminal	V. Trigeminal (V1, V2, V3)
						VI. Abducens
V. Facial, Auditory	V. (first)	V. (first)	V. (first) Vestibulocochlear, (second) Facial	V. (first) Vestibulocochlear, (second) Facial	V. Facial, Auditory	VII. Facial
	Vestibulocochlear, (second) Facial	Vestibulocochlear, (second) Facial				VIII. Vestibulocochlear
VI. Glossopharyngeal, Vagus, Accessory	VI. (first) Glossopharyngeal, (third) Vagus, (second) Accessory	VI. (first) Glossopharyngeal, (third) Vagus, (second) Accessory	VI. (first) Glossopharyngeal, (third) Vagus, (second) Accessory	VI. (first) Glossopharyngeal, (third) Vagus, (second) Accessory	VI. Glossopharyngeal, Vagus, Accessory	IX. Glossopharyngeal
						X. Vagus
						XI. Accessory
VII. Hypoglossal	VII. Hypoglossal	VII. Hypoglossal	VII. Hypoglossal	VII. Hypoglossal	VII. Hypoglossal	XII. Hypoglossal

[a]Both nerves passing through foramen at the back of the eye and going to eye muscles; only first one lies immediately behind the optic nerve, but in the strict sense, because both lie behind the optic nerve, in reality; it is likely that he missed the trochlear nerve.
[b]Importantly, he already recognized those branches.
For details about other Muslim scholars, see text.

Fig. 2.9. Cranial nerves according to Galen, some Muslim scholars and Vesalius, compared to current anatomical terminology.

He described the boundaries and contents of the chest, the diaphragm and lungs (Section 13). The heart was characterised as looking like an 'inverted pine' (Section 14), with the apex pointing inferiorly and its base superiorly. He described a thick membrane that does not adhere to the heart, except at its apex. Like Galen and previous Muslim scholars, he referred to a passage between the right and left ventricles. He illustrated two openings in the right ventricle (inferior vena cava and pulmonary artery) and stated that Galen had said that the first vessel originates in the liver, while Aristotle said it originated in the heart (Leroi, 2014). He then described the stomach as having three layers and the oesophagus as two (Section 15). The small and large intestines were described (Section 16), followed by the location, shape, attachments and cover of the liver (Section 17). He argued that the portal tube originates in the liver and, although it looks like a vessel, it is not really one because it does not contain blood. The spleen and gall bladder were briefly mentioned (Sections 18 and 19). The kidneys (Section 20) are located on both sides of the spine, with the right kidney being higher than the left one. Each kidney has two cavities: one is attached to the great vessel that originates in the liver, and the other (the ureter) descends into the urinary bladder.

He described the urinary bladder (Section 21), its location between the pubic and rectum, its two layers, and the round muscle around its neck (sphincter). Notably, he was mistaken in his description of the course of the ureter through the wall of the bladder. He agreed with Galen's erroneous statement that the ureters run obliquely for some distance in the bladder wall before opening into its cavity and that there is a membrane that acts like a lid, covering the end of the ureters and preventing the backflow of urine. He was seemingly not aware of the work of previous Muslim scholars on this subject (see above). The peritoneum, omentum, and the membrane below the abdominal muscles and above the viscera are described in Section 22, and the penis, testicles, and seminal tubes in Section 23. He described the mammary body as a combination of arteries, veins, nerves, and white glandular flesh (Section 24). The last section (Section 25) covered the ovaries and uterus, its location between the rectum and urinary bladder, and its attachments with loose ligaments. Like Galen, he stated that the uterus has two cavities with a single end (Besteiro and Morales, 1987).

In summary, Ibn Rushd was very interested in anatomy and famously wrote that 'practice of dissection strengthens the faith' (Savage-Smith, 1995; Tbakhi and Amr, 2008). However, like Al-Razi, his descriptions in the anatomy book were brief, with few details. Moreover, and importantly, he followed

Galen's anatomical descriptions even if they had already been proven wrong by previous Muslim scholars. This included the notion of double cavities within the uterus, which was corrected by Ibn Sina's statement that some women have two cavities while others have one, and the ureterovesical junction, which was corrected by Al-Akhawayni when he disproved the presence of a lid at the end of the ureteric opening. Therefore, the anatomical work of Ibn Rushd, the 'prince of science', does not completely match his striking contributions to other fields of science.

Al-Baghdadi, Muwaffaq Al-Din Muhammad Abd Al-Latif ibn Yusuf
1162–1231, Baghdad, Iraq (Arab)

He spent his early life doing intensive studies, mainly in theology and Arabic philology. In 1189–90, he moved to Mosul (a city in northern Iraq), and from there he went to Damascus (Stern, 1962; Leaman, 2006). There he visited Saladin's camp, where the Muslim army was surrounding the city held by the Crusaders, and gained the protection from the minister who gave him a recommendation letter to Cairo.

He lectured in the mosques and studied the writing of ancient scholars, which were available there. When Saladin made peace with the Crusaders (1192), he went to him in Jerusalem and received a pension. After Saladin's death (1193), during the great famine of the Nile River, Al-Baghdadi moved to Egypt and lectured in the Al-Azhar mosque.

He went back to Jerusalem and Damascus in 1207 where he wrote and lectured extensively on medicine. He moved to Aleppo in 1216 and stayed there for about four years before he went to Erzincan (a city in north-eastern Turkey) and got a patron. In 1228 he left Erzincan, travelled to Aleppo, and then went back to Baghdad. He fell sick and died in his native city in 1231 (Stern, 1962; Leaman, 2006).

He travelled broadly but spent most of his life in Aleppo, Cairo and Damascus, which were important intellectual centers in the medieval Islamic world. He studied Arabic grammar, lexicology, and poetry, and was particularly interested in traditional Islamic sciences such as law, jurisprudence, and hadith.

He was also interested in ancient sciences such as mathematics, medicine, and philosophy, having a strong preference for the ancient Greek physicians like Hippocrates, Dioscorides, Rufus of Ephesus, and Galen. However, he also clearly respected some Arab physicians, such as Al-Razi (Joosse and Pormann, 2008).

Al-Baghdadi wrote around one hundred and forty-seven books in different areas, including medicine. Interestingly, while he did not include detailed anatomical descriptions in his medical books, he emphasised the importance of studying anatomy and knowing the parts of the body. He stated: 'He who practises it (phlebotomy) must therefore have a perfect knowledge of the anatomy of the veins, the muscles, and the arteries' (Bonadeo, 2013, p.153). One of his famous books was called *Al-Ifada wa'l-I'tibar*, which had a description of Egypt that included his observations on a famine that occurred there in the year 1200. During this time, Al-Baghdadi was able to observe a huge number of skeletons, from which he concluded that Galen had been incorrect regarding the bones of both the lower jaw and sacrum (Savage-Smith, 1995; 1996; Jobs and Mackenthun, 2011; Bonadeo, 2013). Al-Baghdadi thus established that in human adults the mandible consists of a single bone, rather than two which are separated by a median symphysis. He also showed that the sacrum, as well as the coccyx, include a number of bones but that there are no clear sutures between these bones. That is, both the sacrum and the coccyx are single, complex bones. Unfortunately, this major discovery was largely ignored after his death, both in the Muslim world and in the West, probably because the information was written in a book about the geography of Egypt and not specifically about anatomy (Savage-Smith, 1995, 1996; Jobs and Mackenthun, 2011; Bonadeo, 2013).

Al-Tibb min Al-Kitab wa-Al-Sunna (*Medicine from the Holy Book and the Life of the Prophet*) is an important medical book of Al-Baghdadi. At the end of this book there is a small section about human anatomy, which was based on the Quran and Sunnah. He described the stomach as a nervous hollow body, and called its upper portion the oesophagus and the lower portion the pylorus. In his opinion, the stomach was the place where food was cooked and then moved to the liver to form blood. He also named and described three large intestines and three small intestines, and stated that the total number of intestines, including the stomach, is seven, which is similar to the Prophet's counting. The book briefly described bones and tendons, and the anatomy of the cranial nerves and brain, which, for him, was the origin of sense and movement, sending sensation and movement to every organ. Then he stated that God created a nerve called *Al-Nowri* for the eyes for vision, one for the ears for auditory purposes, one for the nose for smelling, and one for the tongue for tasting (Qala'aji, 1994). Briefly, he described the muscles, spinal cord, foramen magnum, rib cage, brain ventricles, and the sense organs, and explained the importance of each one. He mentioned the Galenic theory of unidirectional blood circulation from the liver to the organs. However, he argued that water accompanies the blood to facilitate its movement, and returns to the liver before travelling through the kidneys and urinary bladder as urine. He provided evidence for his theory, citing the fact of the colour change of urine to red when a woman uses henna (a natural dye used on skin or hair for cosmetic or medical purposes). The journey of water from the organs back to the liver and then to the urinary bladder was described as bi-directional venous blood flow (Qala'aji, 1994; Dalfardi *et al.*, 2014c), and was illustrated for the first time by him. He also described the heart and the great blood vessels that were mentioned in the Quran and Sunnah such as the aorta (*Al-Wateen*) and jugular vein (*Habl Al-Wareed*). Similar to Galen and Ibn Abbas, he stated that the uterus has two cavities, the right one for the male infant and the left for the female infant. He counted three hundred and sixty 'joints' (bones), as mentioned in the Sunnah by the Prophet, together with five hundred and twenty-nine muscles. The hair, nails, and their functions was the last topic of the anatomy part of the book (Qala'aji, 1994).

Al-Baghdadi thus provides a very interesting case study in the history of sciences. Despite focussing only a short section on anatomy, in a book that is mainly based on verses from the Quran and quotes from Sunnah, as well as Galen's works, he corrected important points of Galen's descriptions of human anatomy and physiology. These included Galen's descriptions of the lower jaw, sacrum and coccyx, and of the unidirectional flow for the venous blood circulation.

Ibn Al-Nafis, Abu Al-Hasan, Ala'a Al-Din Ali ibn Abi Al-Hazm Al-Qurashi Al-Dimashqi
1210–1288, Damascus, Syria (Arab)

He studied medicine in a medical college hospital. In 1236 he moved to Egypt and worked in its hospitals, becoming chief of physicians and the Sultan's (ruler's) personal physician (Leaman, 2006; Al-Ghazal, 2007; West, 2008).

He learned jurisprudence, literature, and theology and was a renowned expert in the Shafi'i school of jurisprudence as well as a reputed physician (Al-Ghazal, 2007). When he died in 1288, he donated his house, library, and clinic to the Mansuri Hospital (Al-Ghazal, 2007).

He wrote many medical books, some of them including his commentaries on Hippocrates and others on Ibn Sina and Hunayn ibn Ishaq (a famous Christian scholar). One of Ibn Al-Nafis' medical books, *Al-Kitab Al-Shamil fi Al-Tibb*

(Comprehensive Book in Medicine), was planned to be an encyclopedia with 300 volumes, but he finished only 80 volumes before his death (Leaman, 2006; Al-Ghazal, 2007). Another famous book, *Mujaz Al-Qanun* (*The Summary of the Canon*), summarized the *Canon* of Ibn Sina, excluding the anatomy and physiology parts. These parts were included in another book called *The Commentary on Anatomy in Avicenna's Canon* in which he criticised Ibn Sina for dividing his anatomy sections in different books of Al-Qanun (Savage-Smith, 1995). He was only twenty-nine when he wrote this book, which is considered his most important work, as it included his ground-breaking views on the pulmonary circulation and heart (West, 2008). He stated in the introduction of his *Commentary* that he would relay the descriptions of the internal organs on the knowledge of previous scholars who practised dissection like Galen, but not the wrong theories.

The Commentary on Anatomy in Avicenna's Canon is considered Ibn Al-Nafis' most important work and included his ground-breaking views on pulmonary circulation and the heart (West, 2008). Ibn Al-Nafis stated in his introduction that his descriptions of the internal organs were based on the knowledge of previous scholars, such as Galen, who practised dissection but would not include any incorrect theories by those authors. The book had two parts. The first concerns the internal organs and has five chapters about bones (in thirty sections), muscles (twenty-nine sections), nerves (in six sections), arteries (five sections), and veins (five sections). The second part has twenty chapters about the brain, eye, ear, nose, mouth and tongue, pharynx, larynx and lungs, heart, breast, oesophagus and stomach, liver, gallbladder, spleen, intestines, kidneys, urinary bladder, testis and seminal tubes, penis, uterus, and, finally, about the delivery of a foetus. The book also has drawings of cranial sutures similar to those shown in **Fig. 2.3**, as well as of the upper jaw and abdominal muscles.

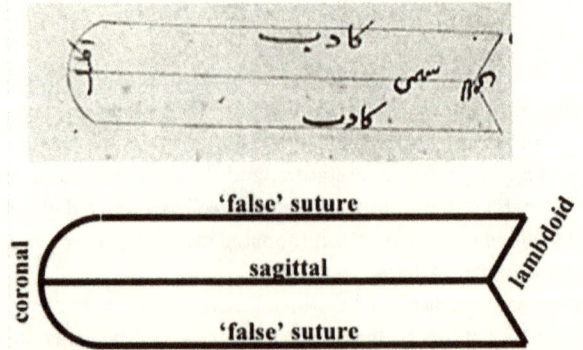

Fig. 2.3. A diagram that can be commonly found in the books of Muslim scholars showing the cranial sutures.

He divided the joints in the bone section into three types: the joint between two bones, such as the head bones (which is an immovable joint, in modern terminology); the joint between two cartilages, such as those in the long bones of the limbs; and the joint between the bone and cartilage, like the one in the sternum. He stated that Galen did not consider the first type as a joint. He described and named the cranium bones, the cranium sutures, and the bones of the upper and lower jaws, reporting two bones in the lower jaw. It seems that he was, thus, unaware that Al-Baghdadi had disproved the idea in 1203 (Ibn Al-Nafis, unknown date).

In the second part, which was about the muscular system, Ibn Al-Nafis followed Ibn Sina's arrangements and explained his descriptions of each region in detail, added some comments, and clarified the meaning of some sentences. In his descriptions of the cranial nerves, he stated that the first pair of nerves (the optic nerves) meet at the optic chiasm, and that Ibn Sina said that they cross each other, i.e. the right nerve goes to the left eye and the left nerve goes to the right eye. However, he also said that Galen considered that they only meet and then split without crossing, which was the common theory at that time. Furthermore, he wrote that some anatomists described the third and fourth pairs (branches of the trigeminal nerve) as one nerve, but Galen described them as two nerves that mixed together at the beginning and then split. Regarding the fifth cranial nerve (vestibulocochlear and facial), he wrote that Ibn Sina might be mistaken in his description, because he considered that each nerve of the fifth pair was actually a double and divided into two nerves. He then explained the difference between arteries and pulmonary veins. Importantly, in this book, the pulmonary circulation was described for the first time in much detail. As explained above, this circulation was not described by Galen, and only Al-Akhawayni had provided some accurate details about it. He contradicted Galen's reports on the presence of a pathway of 'invisible pores' or a visible hole between the right and left cavities, and stated that blood

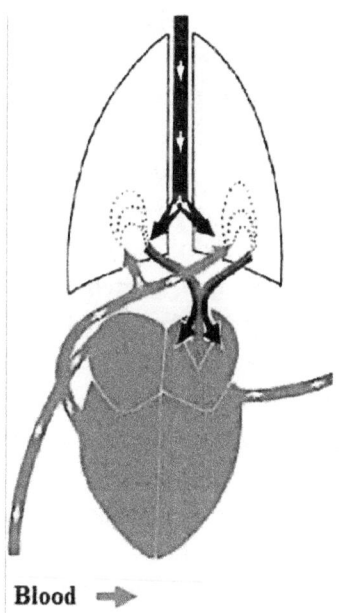

Fig. 2.5. Illustration of the pulmonary circulation according to Ibn al-Nafis.

moves to the lung through the vena arteriosa (pulmonary arteries). There, it mixes with air and is filtered before moving back to the left cavity via the arteria venosa (pulmonary vein) (**Fig. 2.5**; Ibn Al-Nafis, unknown date).

Ibn Al-Nafis also assumed that nutrients for the heart were obtained from the coronary arteries and disproved Ibn Sina's statement that the arteria venosa (pulmonary vein) nourishes the lungs because, for him, this vessel carried the blood to the heart and not to the lungs (see image). On the other hand, he did not challenge Ibn Sina's inaccurate descriptions of the ascending aorta. In the part about the venous system, he stated that Galen described the heart as the origin of the arteries, the liver as the origin of the veins, and the brain and spinal cord as the origin of the nerves, while Aristotle considered arteries and veins to originate in the heart (Leroi, 2014). Ibn Sina had argued that both these Greek authors were partially correct, but Ibn Al-Nafis opposed them and defended the theory that the blood vessels do not originate from other organs, but form, like any other organ, without a specific origin.

In his description of the brain, Ibn Al-Nafis characterised its four ventricles and rejected the notion of empty cavities supposedly filled with spirit (pneuma), thought to be responsible for faculties such as sensation, imagination, and memory, as he could not observe any evidence of them. He defined the covering of the brain, the tentorium, the longitudinal cerebral fissure, the brain's venous system, and the circle of Willis. His descriptions of the eye were similar to previous scholars in possessing three humours and seven layers. However, he stated that the glacial humour (the lens) is not responsible for vision because it was covered with a dark layer (the iris), and, according to him, it was this dark layer that received vision (Ibn Al-Nafis, unknown date).

In the section about the lungs, he stated that they receive nourishment through the vena arteriosa (pulmonary arteries), and that the arteria venosa (pulmonary veins) carry blood and air to the left cavity of the heart. Furthermore, he wrote that the descending aorta passes behind the diaphragm at the level of C12, and not through the diaphragm as the oesophagus does; a theory proposed by Ibn Sina. He illustrated the heart and stated that there was no bone at its base, and that the only bones in this area were the ones forming the rib cage. He then completed his description of the pulmonary circulation (**Fig. 2.5**) and wrote: 'He (Ibn Sina) said it (the heart) has three ventricles, but this is not a true statement. Indeed, the heart has only two ventricles, one of them filled with blood, on the right side, and the other filled with pneuma on the left. There is definitely no passage between these two, for otherwise the blood would pass to the place of the pneuma and would degrade its essence and (furthermore) dissection (*tashrih*) refutes what they said, for the septum (*hajiz*) between the two ventricles is much thicker than elsewhere' (Savage-Smith, 1995, p102). This direct reference to dissection is very significant (see below).

While describing the oesophagus and stomach, he correctly identified the position of the beginning of the stomach and stated that it was located behind the xiphoid process of the sternum and not at the level of C12, as was commonly accepted by physicians of his time. He agreed with Galen and others about the formation of the blood in the liver from digested food, 'Chyle and Chyme', that came from the stomach. His description of the ureters' course through the wall of the urinary bladder was similar to Ibn Sina's, i.e., similar to current knowledge and different from Galen's mistaken descriptions. Galen and Ibn Abbas described a single tube for the delivery of both urine and semen in the penis. However, Ibn Al-Nafis and Ibn Sina described three tubes inside the penis; one for the urine, one for the semen, and another for *al-wade* (for Ibn Sina) or *al-mathe* (for Ibn Al-Nafis). These three tubes joined together at the end of the penis. Ibn Al-Nafis described *al-mathe* as a thin fluid produced before semen at the beginning of sexual intercourse in a gland (prostate gland) which was located at the beginning of the third tube. He also described *al-wade* as a fluid that protected the urinary tube and was produced from a gland located at the beginning of the urinary tube (bulbourethral glands). It is important to note that the prostate gland and bulbourethral gland were neglected by scholars who preceded Ibn Al-Nafis.

In summary, in *The Commentary on the Canon,* Ibn Al-Nafis quoted and explained all of Ibn Sina's descriptions and some of Galen's too, and commented on them, being either in agreement or disagreement. This book has more details, descriptions, and illustrations than the books of other Muslim scholars because it gathered the anatomical descriptions of many anatomists. Importantly, despite a few mistakes, most of Ibn Al-Nafis' anatomical descriptions are similar to our current anatomical knowledge, his major contribution being the accurate description of pulmonary circulation. Unfortunately, this description of pulmonary circulation was neglected by the Muslim and Western scholars who succeeded him, before it was re-discovered three centuries later (Al-Ghazal, 2007; Savage-Smith, 1996).

Mansur ibn Ilyas, Mansur ibn Mohammad ibn Ahmad ibn Yousef ibn Ilyas
1380–1409, Shiraz, Timurid Persia

He was trained under one of his family members, attended traditional schools in Fars, and visited many cities including Tabriz (a city in northwest Iran), where he enriched his knowledge (Khalili et al., 2010).

He wrote numerous medical books, one of them called *Kifaya-yi Mujahidiya* or *Kifaya-yi Mansuri*, which included a short anatomical section based on Galen's dissection (Shoja and Tubbs, 2007; Zarshenas et al., 2014). After that, he wrote a book called *Tasrih-i Mansuri* (*Mansur's Anatomy*), also known as *Tashrih-i Badan-i Insan* (*Human Anatomy*), which had coloured illustrations (Shoja and Tubbs, 2007; Khalili et al., 2010). This book was dedicated to Prince Pir Mohammad Bahador, the grandson of Tamerlane, the ruler of the Persian province of Fars from 1394 to 1409 (Shoja and Tubbs, 2007; Khalili et al., 2010; Ziaee, 2014).

Mansur ibn Ilyas wrote a book called *Tasrih-i Mansuri* (*Mansur's Anatomy*), also known as *Tashrih-i Badan-i Insan* (*Human Anatomy*), which had coloured illustrations (Shoja and Tubbs, 2007; Khalili *et al.*, 2010). *Mansur's Anatomy* was written in the Persian language and in a systematic manner. It had an introduction with remarks about Aristotle, Hippocrates, Galen, Al-Razi, Ibn Sina, and the prophetic tradition, followed by five sections on bones, nerves, muscles, veins and arteries, and a concluding chapter on complex organs and foetal development (Contadini, 2007; Khalili *et al.*, 2010).

In the introduction, Mansur discussed which organ was the first to be differentiated in the uterus (Ziaee, 2014) and stated that, as the heart is the structure that surrounds the spirit, it should, in theory, be the first organ to develop (Khalili *et al.*, 2010). The skeleton (Section 1) contains simple drawings of the upper jaw, cranial bones, and sutures (**Fig. 2.6**; compare with **Fig. 2.3**; Zarshenas *et al.*, 2016; Ziaee, 2014). The nervous system is described in Section 2, where he reports that nerves originate in the brain and must be hollow and that they also convey a spirit. Despite some errors, Mansur described the gross anatomy of the nervous system well. He stated that there were seven cranial nerves and considered the filamentum terminale to be a single nerve. As such, he wrote that there were thirty-one pairs of spinal nerves and one odd one (Zarshenas *et al.*, 2016; Ziaee, 2014).

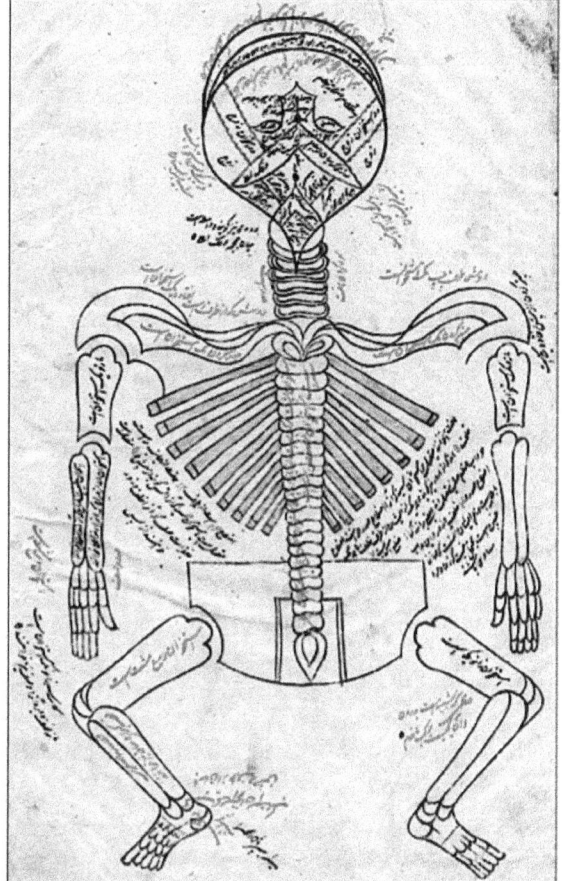

Fig. 2.6. Figure of the skeletal system illustrated from Mansur's Anatomy.

The muscular system (Section 3; **Fig. 2.7**) was the least comprehensive section in the whole book, including, as it did, mostly comments on the gross structure of muscles, the varieties of muscles as they appear to the naked eye, and the number of muscles in the body (Zarshenas et al., 2016; Ziaee, 2014). In Section 4, he stated that venous blood carries hepatic blood and the natural spirit, and starts its journey in the liver, just as arterial blood starts its journey from the heart, following Galen's misconception. As with Ibn Sina, Mansur considered that the arteria venosa (pulmonary vein) nourishes the heart and lungs. As noted above, this was contradicted by Ibn Al-Nafis. As with Galen and Ibn Sina, Mansur erroneously thought that the venous blood flows within the heart, and that the heart has three ventricles and is strengthened by a special bone (Ziaee, 2014).

The arteries—their origins in the heart, their pulsation in time with the heart, and the number and direction of layers—were reported in Section 5. Mansur described two arteries that originate in the left cardiac ventricle—that is, the aorta and venous arteries—and stated that they travel to the lung, where they then branch. He also described the coronary arteries, i.e., the ascending aorta, and descending aorta (Khalili *et al.*, 2010). The last chapter comprised the compound organs, in which he defined organs as structures of which the smallest part exactly resembles the whole. Thus, a portion of a bone is still a bone, and a branch of an artery has still to be called an artery. A compound organ is one that cannot be subdivided. Therefore, the heart can be divided up into ventricles and auricles, but none of these alone can still be called a heart (Ziaee, 2014). In the conclusion, he described the location and relationships of the heart, referring to the presence of a hole in the wall between the right and left cardiac ventricles which allowed blood to move from the right to the left side (Khalili *et al.*, 2010)

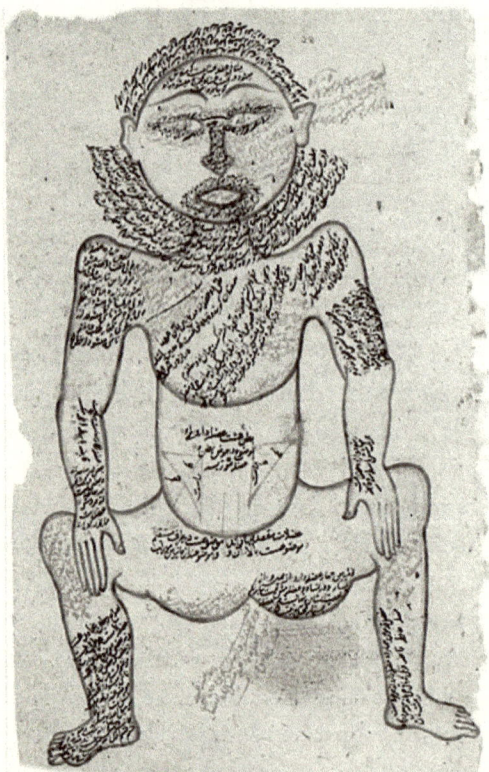

Fig. 2.7. Figure of the muscular system illustrated from Mansur's Anatomy.

The major contribution of *Mansur's Anatomy* is that it contained six full-length coloured images (Zarshenas et al., 2016); however, there is some controversy regarding the origin of these illustrations (Shoja and Tubbs, 2007). Contadini (2007) described them in her book, saying that the first and second images, which depict the skeleton (**Fig. 2.6**) and nervous system, are viewed from behind, with the head hyper-extended so that the mouth is drawn at the top of the page and the palm turned towards the viewer. She also stated that the image of the muscle (**Fig. 2.7**) is shown from the front with captions describing the abdominal muscles as triangles in the lower abdomen. This was similar to the Arabic drawing of these muscles. The arterial and venous systems (**Fig. 2.8**) were also illustrated from the front, with an indication of the internal organs and identification of the structures. Additional figures show the compound organs and a foetus as a mature male in breech position. The figures were labelled in a mixture of Arabic and Persian languages, and some authors have said that the last figure was the only original contribution made by Mansur, the others being taken from earlier sources. For instance, similar anatomical drawings were found in the late 19th century which were thought to have been produced in Europe in the middle of the 12th century (Contadini, 2007). Savage-Smith (1996) also stated that the illustrations accompanying Mansur's book appear to maintain the Greco-Roman tradition in anatomy, and they are related to Latin anatomical diagrams from the 12th century. However, it is still unknown in what form, or by what means, such full-length anatomical diagrams were available to Mansur in his time (Contadini, 2007). In summary, it can be said that, although Mansur seemingly did not make a major

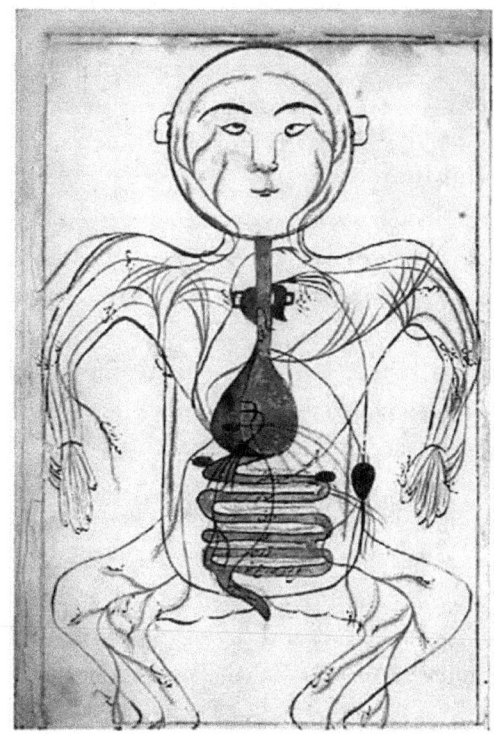

Fig. 2.8. Figure of the veins illustrated from Mansur's Anatomy.

contribution towards discoveries in human anatomy, his compilation of full-length diagrams in a single volume may have provided a valuable source of information for anatomists, physicians and other scholars of that time.Ω

General remarks on the anatomical discoveries of pre-Vesalius Muslim scholars

Most accounts on the history of the anatomical sciences move rapidly from the Greco-Roman period to the European Renaissance. They ignore the scientific contributions of Muslim scholars in the Islamic Golden Age that bridged the gap between Eastern and Western cultures. However, our work clearly shows that while in some respects Muslim scholars followed Galen's anatomical observations, many of them produced their own original contributions to anatomy, particularly concerning osteology, the heart and pulmonary circulation, the circle of Willis, the relationship between the ureters and the urinary bladder, and the eye, among many others. The anatomical knowledge accumulated during the Islamic Golden Age, and compiled by scholars such as Ibn Al-Nafis, is clearly more in keeping with current knowledge than the studies of Galen.

The source of these original anatomical discoveries by Muslim scholars could have been human or animal dissections when they practised surgery, or observed the cadavers of humans who had died accidentally, as in the case of Al-Baghdadi. It is often assumed that Muslim scholars did not practise dissection because it is prohibited under Islamic Law. However, as noted above, there is no explicit statement in Islamic religious texts that either supports or opposes the practice of dissection. Importantly, the impact of other factors—for instance, burial customs and beliefs about the dead body—need to be further explored. Based on details provided by some pre-Vesalius Muslim scholars, as well as the fact that a number of them explicitly referred to the importance of dissection in their works (e.g., Al-Razi, Ibn Rushd and Ibn Al-Nafis), it seems likely that several of these scholars conducted some kind of human dissection.

Taking into account everything we have read in this chapter, it seems particularly puzzling that, regarding the muscular system specifically, there were apparently no advances by either Muslim or European scholars for more than a millennium after Galen. The descriptions of prominent Muslim authors—at least the earlier ones such as Ibn Abbas and Ibn Sina—basically repeated Galen's erroneous descriptions of 'human anatomy' based on his dissections on animals such as monkeys. How is it possible that for more than a millennium nobody dissected, or even observed during surgeries or autopsies, the structure and form of human muscles? This is perhaps one of the most intriguing questions, not only for the history of anatomy, but also for the history of biology, and the sciences, in general. It is even more puzzling when one considers the evidence of human dissections being performed by Greek scholars as early as 300 BC or even earlier; a practice that seems to have been discontinued during the time of the Roman Empire (Singer, 1957: 14, 28, 38). One reason for this might be that there was a more pragmatic view of sciences within the Roman Empire, as well as in the European Middle Ages that followed it. According to this view, knowledge gained by manual labour was strongly stigmatised in elite circles (Daston and Park, 2001: 118), and muscle dissections were discouraged. In turn, in the more physiologically or medically driven context of Muslim pre-Vesalius science, knowledge of muscle anatomy was probably not seen as particularly important per se, compared to the anatomy of organs, such as the heart, lungs and liver. That is, it is likely that Muslim scholars carefully dissected some parts of the body, such as the eyes or the heart, but did not dissect muscle tissue in so much detail.

A topic we plan to address in the future concerns the influence of progress made by Muslim scholars on subsequent Western scholars, and, thus, to modern anatomical knowledge in the West. The dominant view is that in the 15th and 16th centuries, Westerners realised that the best way to contribute to the revival of anatomy was to study ancient Greek texts directly, in order to not be 'corrupted' by Muslim texts (Singer, 1957, 159). This seems to be supported by the fact that modern anatomical terminology is mainly based on Greek and Latin terms. This idea is additionally endorsed by a very interesting and intriguing fact: that there is direct evidence that some major discoveries made by Muslim scholars were neglected, or at least forgotten with time, not only by Westerners, but also other Muslim scholars. Some cases might be easier to understand, for example, the osteological advances made by Al-Baghdadi might have been neglected because he published them in a book dealing mainly with geography. However, other cases, e.g., concerning Ibn Al-Nafis' comprehensive description of the pulmonary circulation—some three hundred and seventy-four years before Harvey's famous work—seem more puzzling.

However, taking into account that the books written by Muslim scholars were widely used by Westerners in the Middle Ages, that some of them were highly influential until the 18th century, and that Muslim scholars made notable anatomical advances relative to Galen, it seems likely that these advances did have some influence in the subsequent development of anatomy in the West. This idea is supported by the fact that, as recognised by historians of science, a significant portion of the current anatomical terminology is actually based on Arabic terms e.g., nucha, basilica, retina, saphena, sesamoid (Singer, 1957). And, given that the anatomical terminology was actually revised in 1895—to render it into 'correct' Latin—as the *Basiliensia Nomina Anatomica* (Sakai, 2007), there were possibly many more Arabic terms in use before then.

In conclusion, one can say that there have been historical biases in Western accounts on the history of biology and anatomy that have not allowed us to fully understand the true complexity and geographical diversity of the history of discoveries in human anatomy. This lack of understanding contributes to the propagation of unfortunate stereotypes and, importantly, does not allow a proper sense of the history of anatomical knowledge, biology, or science, as a whole. In fact, it is important to note that there is increasing evidence that, apart from pre-Vesalius Muslim scholars, non-Muslim authors, such as Moses Ben Maimon (Maimonides) and a number of Chinese scholars, were making relevant discoveries about human anatomy well before Vesalius was born (Cowdry, 1921; Stroumsa, 2012; Shaw, 2014; Standring, 2016), and even before the birth of Galen, in the case of Ancient China (Shaw *et al.*, in press). The present work, therefore, paves the way for much needed future works on these topics. This includes, for instance, a detailed literature review about the development of anatomy in Europe and its specific sources, as well as the change in anatomical terminology, from the 15th century to the present time. Specifically, it is crucial to investigate if there were, or were not, major contributions from non-Greek and non-Western scholars, in order to increase our knowledge of the history of anatomical knowledge, biology, and sciences as a whole, and also to be able to discuss a broader range of subjects, such as the transmission and neglect of knowledge and, thus, better understand our own way of thinking, including our own biases and prejudices.

CHAPTER 3

UNTOLD STORIES: MUSLIM SCHOLARS AND EVOLUTIONARY IDEAS

"I died as a mineral and became a plant, I died as a plant and rose to animal, I died as an animal and I was Man. Why should I fear? When was I less by dying?"
(Rumi)

Religion, creationism, human evolution, and stereotypes

The view that Muslims are, in general, reticent about the idea that all species—especially our own—have been changing over time is based on factual quantitative analyses, which is often exaggerated to reinforce certain narratives. This exaggeration is carried out, not only by many non-Muslims, including a substantial number of Western lay people and scholars, but also by fundamentalist Muslim groups such as ISIS (Islamic State of Iraq and Syria), also known as ISIL (Islamic State of Iraq and the Levant), and Al-Qaeda. Why is this? Well, because it is appealing for, and serves the narratives of, these non-Muslim *and* Muslim groups to say that science is incompatible with Islam. Non-Muslims can use such narratives to criticise *all* Muslims and their 'fanaticism' and 'backwardness' and thus perceive themselves and their societies as the 'pinnacle of evolution' (see **Chapter 1**). In turn, ISIS and Al-Qaeda can defend their false narratives that 'true' Muslims always reject science or Western ideas, not knowing, of course, that it was not only Westerners who had such ideas.

In many Islamic communities, the theory of evolution is a controversial subject due to its perceived conflict with Islamic teachings (Asghar *et al.* 2007; Hameed, 2008). During the 20th century, evolution came to be associated with provocative ideas, including atheism, materialism, colonialism and imperialism, leading Muslim scholars and their communities to reject it

(Dajani, 2015). Interestingly, several Muslim philosophers and public figures between the late 19th and early 20th centuries, such as Ahmed Medhat, were vocal in their support of evolutionary ideas. Indeed, some authors have gone so far as to say that 'up to the modern era [early 20th century], they [most Muslim intellectuals] by and large accepted biological evolution and even welcomed it, as long as it did not present itself in purely materialistic, atheist garb, even though the question of human evolution did often constitute a sore point; nowadays, however, the rejection of Darwinism is nearly unanimous (in the Muslim world)' (Guessoum, 2011). It is, therefore, not surprising that the contributions of pre-Darwinian Muslim thinkers to the history of evolutionary thought remains largely unrecognised today, even within the Muslim world. In fact, evolution is such a sensitive topic in some parts of the Muslim world that *fatwas* (legal opinions or decrees by an Islamic religious leader) have been issued on it (Guessoum, 2011). As noted above, this is part of an attack on what are, wrongly, perceived as 'Western' ideas. In fact, one should keep in mind that the name of another Islamic fundamentalist group, Boko Haram, is usually translated as 'Western education is forbidden'.

Paradoxically, the erroneous narratives defended by such fundamentalist Muslim groups are providing fuel to anti-Muslims by minimising, or, in some cases, neglecting, the amazing scientific discoveries and profound discussions about the natural world carried out by Muslims in the past (particularly during the Golden Islamic Age), which they so much revere. To complicate things even more, it is also true—and often neglected in the West, particularly by those who criticise current Muslim scholars for not accepting that humans are derived from other primates—that one of the major causes leading to the scepticism about Darwin and his works within the Muslim world, is that they were perceived as a symbol of Western 'superiority', imperialism, and colonialism. This viewpoint was put forward in Muzaffar Iqbal's 2007 book, *Science and Islam*. As noted above, it is also true that Darwin and his works were—and still are, in great part—celebrated, used, and even revered in the West precisely because some Westerners, including certain scholars, perceive themselves as the 'winners' of a double victory as symbolised by Darwinism: that of Western science and knowledge, and against 'irrational' beliefs. The latter point is ironic because, as we have seen above, many of the inaccurate ethnocentric ideas that Darwin wrote as 'scientific facts' in his books about human evolution were indeed based on his *belief* in biased narratives and prejudices, rather than on what he truly saw amongst the non-European peoples he encountered during his Beagle voyage.

In other words, the intolerant approach of many Muslims, and particularly Islamic fundamentalist groups, concerning what they call 'Western' science, and evolutionary biology in particular, over the last one hundred and fifty years, somewhat parallels the extremely biased way in which Westerners have interpreted the history and contributions of Muslims. In the past, some Western researchers did acknowledge the transmutation ideas of various pre-Darwinian Muslim scholars, but today one struggles to find any mention of those ideas in history textbooks and specialised papers on these topics. For instance, John William Draper (1812–1883), a renowned 19th century scientist and contemporary of Darwin, was critical of Western science's dismissiveness towards medieval Muslim thinkers (Shanavas, 2010). He wrote: 'I have to deplore the systematic manner in which the literature of Europe has contrived to put out of sight our scientific obligations to the Muhammadans... Injustice founded on religious rancour and national conceit cannot be perpetuated forever' (Draper, 1875, 1876). Draper recognised the transmutation ideas of some Muslim scholars and, in *The History of the Conflict Between Religion and Science*, he wrote: '[Christian] (t)heological authorities were... constrained to look with disfavour on any attempt to carry back the origin of the earth to an epoch indefinitely remote, and on the Muhammadan theory of evolution which declared that human beings developed over a long period of time from lower forms of life to their present condition.'

Draper further elaborated on a typical attitude of many Western scholars: 'Sometimes, not without surprise, we meet with ideas with which we flatter ourselves with having originated in our own times. Thus our modern doctrine of evolution and development were taught in their [Muslim] schools. In fact, they carried them much farther than we are disposed to do, extending them even to inorganic and mineral things.' Because their ideas are often deemed to be more religious or philosophical than scientific, medieval Muslim scholars such as Tusi and Ibn Tufayl, have neither been credited by Western scholars for having transmutation ideas, nor have they received due attention for their ideas. As Draper pointed out, this appears to be a reflection of historical prejudice, because scientific history has been shown to recognise theories far more abstract and less plausible than those of the scholars discussed in this paper (see Gould's 2002 review of the history of transmutation ideas).

It is striking how current Western textbooks differ from Draper's statements. The idea repeatedly defended in such textbooks, including those specifically dedicated to the history of biology, is that 'after Galen we encounter no biological activity for centuries', or that the 'Dark Ages in science', extended 'from the death of Galen until the 13th century' (Singer, 1959). As noted above, these scholars recognise the existence of a Golden Islamic Age, but they mainly associate it with some kind of 'innovation' in the fields of mathematics and medicine and not in comparative anatomy, as we have seen in Chapter 2, or concerning transmutation ideas, which will be discussed in this chapter. To illustrate how Western biologists and historians of science tend to refer to this issue, we can cite a passage by one of the most respected biologists of recent times, Ernst Mayr, from his highly influential 1982 book about the history of biology, *The Growth of Biological Thought*: 'The Arabs, so far as I can determine, made no important contributions to biology—this is even true for two Arab scholars, Avicenna... and Aberrhos (Averroes)... who showed a particular interest in biological matters; it was, however, through Arab translations that Aristotle again became known to the Western world—this was perhaps the greatest contribution that the Arabs made to the history of biology.' Again, the narrative is about Muslims as passive players: they can only translate, not innovate, and Mayr is referring not only to transmutation ideas, but also to biology. That is, the study of life. This is very odd to say the least, because it is well known that scholars, such as Avicenna, wrote some of the most complete and extensive texts on medicine—a field concerned with life, death, disease, and the human body and its physiology.

We can see here an interactive cycle, whereby negative reactions against evolutionary ideas by Muslim scholars were, in part, due to Western biases and prejudices, and the 'inferiority' of 'others'. Because of this, many Muslim scholars, politicians and religious leaders started to oppose anything resembling transmutation ideas. This has been particularly true in the last few decades and, because of it, some Western scholars, such as Mayr, extrapolating the 'always' from the 'now', concluded that Muslim scholars have never provided any valuable contribution to biology as a whole; something that is obviously and factually inaccurate. This is a typical cycle of misunderstanding, bias, and prejudice, in which the 'us' *versus* 'them' becomes more important than the reality, and in which, accordingly, the false narratives constructed by the more extremist and fundamentalist members of each of the two groups gain predominance over more moderate and realistic narratives.

In fact, the two main reasons discussed above—that 'transmutation was always rejected by Muslim scholars', which is not true, as we will see, and 'Darwin was used as a symbol of Western supremacy and imperialism', which is, in great part, true—are increasingly being used by fundamentalist Muslim scholars to reject and actively fight against transmutation ideas. As summarised by Saouma B. BouJaoude *et al.*, in a 2011 paper entitled, *Biology Professors' and Teachers' Positions Regarding Biological Evolution and Evolution Education in a Middle Eastern Society*:

> The social controversy over evolutionary science and its teaching is likely to be carried to the classroom because students and teachers are influenced by their cultures and societies... and the situation could grow even more dire... evolution may be becoming increasingly controversial in Muslim societies where ideological religious arguments against the theory are popularized in many books, articles, and Internet sites on the topic produced for the general public. Research in North America has shown that significant numbers of college students reject evolution [see also **Fig. 3.1**]. Some recent studies conducted with college and secondary students in predominantly Muslim societies suggest a similar trend. Research in Lebanon has shown that, while most university biology students understand the principles of evolution, approximately 50% of them reject the conclusions of evolutionary science. This rejection, however, may well be related to students' religious affiliations as evidenced by the fact that while 82% of Christian university students accepted evolution, only 35% of Muslims did so. Additionally, Dagher and BouJaoude found that university students: (1) accepted evolutionary ideas using arguments from a scientific or religious-reconciliation perspective; (2) did not accept evolutionary ideas presenting arguments from a religious or anti-evolution perspective; (3) reinterpreted the theory presenting arguments from a compromise perspective; and (4) were neutral, reflecting either a non-committed or a confused perspective.

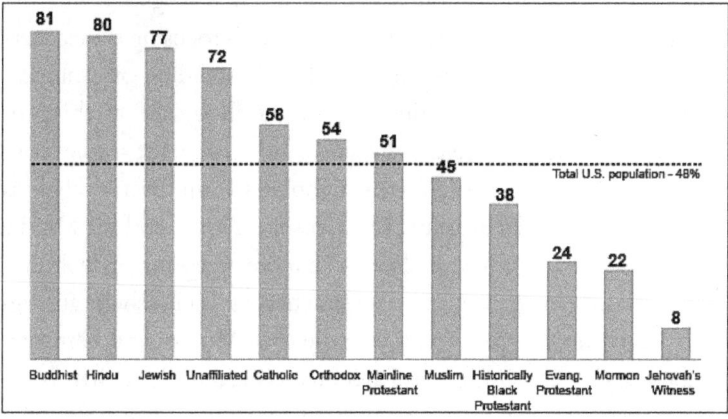

Fig. 3.1. Percentage of people, from different U.S. 'groups', that agree that evolution is the best explanation for the origins of human life on earth, according to a Pew Research Center 2009 study.

In another study with university biology students in Lebanon who completed a course on evolutionary science, Hokayem and BouJaoude found that students' positions ranged from complete acceptance to complete rejection of evolution. In a study conducted in Lebanon and Egypt with secondary-level students, BouJaoude and Kamel found that secondary-level students thought that evolution is both a fact and a theory, but also that evolution is a theory because it is not yet 'proven' scientifically. The latter finding indicates that these students held a common fundamental misconception related to the nature of science [that is, in science hypotheses can only be contradicted or supported—never proven]. BouJaoude and Kamel also found that a higher percentage of Lebanese Muslim students than Christian or Druze students thought that their religion teaches that the first life and humans on Planet Earth were created by God, not gradually but suddenly in their present human form; accurate science includes religious explanations; evolution is best learned from the holy book of their religion; biology classes should include their religion's explanations of human and animal history on Earth; God created human beings pretty much in their present form at one time within the last 10,000 years or so; and their religion influences how they think about evolution.

Alternatively, results regarding Egyptian students' conceptions of the relationship between science and religion showed that Muslim and Christian Egyptian students had similar misconceptions about evolution except that a higher percentage of Muslim than Christian students thought that biology classes should include the religious explanation of human and animal history and that religion influences how they think about science. Related research focusing on teachers suggests that their religious beliefs and understanding of evolution may influence their instructional decisions about the teaching of evolution in their classrooms... Finally, a series of recent studies conducted among teachers of various disciplines from diverse religious backgrounds in different countries in Europe, Africa, and the Middle East revealed that biology teachers were more accepting of evolution than were teachers of language or elementary school, atheist and agnostic teachers accepted evolution more than Muslim and Christian teachers, and Christian teachers accepted evolution more than Muslim teachers. Importantly, they showed that there were differences within Christians; with Orthodox teachers—who came mainly from Romania or Cyprus—more opposed to evolution than Catholic or Protestant teachers. Moreover, they found that there was widespread rejection of evolution in countries with Muslim majorities. Clément and Quessada elaborated on these findings, noting that teachers who identified themselves as Christians in Northern European countries, such as France, Germany, and Finland, were more accepting of evolution than Christians from other countries such as Cyprus and Lebanon. Quessada, Munoz, and Clément found that irrespective of the level they taught or the country they came from, teachers with longer training were more accepting of evolution.

BouJaoude's summary is very useful. It highlights that today (and it is important to emphasise the *today*), Muslim scholars and students tend to be unwilling to accept biological evolution, and in particular human evolution, than other religious or non-religious groups, including those in the U.S. However, when we analyse numbers in the U.S. in a more profound way, we see that the perception that Muslims are different from other groups when it comes to the acceptance of biological evolution is also exaggerated. A 2009 *Pew Research Center* report entitled *Religious Differences on the Question of Evolution*, showed that about half of U.S. Muslims agreed that evolution is the best explanation for the origins of human life on Earth (**Fig. 3.1**). This is clearly different from such stereotypical comments as: 'no Muslim accepts human evolution': and is a difference of 45%, that is, from 0% to almost 50%. Moreover, this average was only slightly lower than the percentage of the total U.S. population (48%) who accepted biological evolution (**Fig. 3.1**). In fact, the average for some U.S. Christian groups, such as, 'Historically Black Protestants' and 'Evangelical Protestants, was lower than that of U.S. Muslims (**Fig. 3.1**).

Also, in a *Pew Research Center* report published in 2003 entitled *The World's Muslims: Religion, Politics and Society*, it was reported that:

> Many Muslims around the world believe in evolution. In 13 of the 22 countries where the question was asked, at least half say humans and other living things have evolved over time. By contrast, in just four countries do at least half say that humans have remained in their present form since the beginning of time. In Southern and Eastern Europe, a majority of Muslims in Albania (62%) and Russia (58%) believe in evolution. But Muslims are divided in Bosnia-Herzegovina

(50% believe humans have evolved, while 45% take the opposite view) and Kosovo (34% vs. 40%). In four of the Central Asian countries surveyed, more than half of Muslims say they believe in evolution, including nearly eight-in-ten in Kazakhstan (79%). In Tajikistan and Turkey, by contrast, the predominant view is that humans have remained in their present form since the beginning of time (55% and 49%, respectively). At least six-in-ten Muslims in Lebanon (78%), the Palestinian territories (67%) and Morocco (63%) think humans and other living things have evolved over time, but Jordanian and Tunisian Muslims are more divided on the issue. About half in Jordan (52%) believe in evolution, while 47% say humans have always existed in their present form. And in Tunisia, 45% say humans have evolved, 36% say they have always existed in their present form, and 19% are unsure. Iraq is the only country surveyed in the Middle East/North Africa region where a majority rejects the theory of evolution (67%).

Muslims' views on evolution vary in South Asia and Southeast Asia. Muslims in Thailand (55%) and Bangladesh (54%) tend to accept that humans have evolved over time. But Muslims in Malaysia and Pakistan are divided: roughly four-in-ten Malaysian Muslims (37%) believe in evolution, while 45% say humans have always existed in their present form. In Pakistan, 30% think humans have evolved, while 38% disagree and 32% say that they do not know. In Afghanistan and Indonesia, the prevailing view is that humans and living things have remained in their present form since the beginning of time (62% and 55%, respectively). In countries surveyed in Southern and Eastern Europe, more religiously observant Muslims are less likely to believe in evolution. In Russia, for example, 41% of Muslims who pray several times a day believe in evolution, compared with 66% of those who pray less frequently. Significant gaps also appear between more and less devout Muslims in Bosnia-Herzegovina (-19 percentage points) and Kosovo (-14). Views on evolution do not differ significantly by religious commitment in the other regions surveyed.

Moreover, it is important to note that some of the Muslim scholars who reject biological evolution embraced a viewpoint similar to that of Western Christian creationists, thereby 'transferring the Western war between science and religion to Islam' (Dajani, 2015)—a war that is being fought, within the Muslim world, between the more fundamentalist and the more moderate scholars. The role of moderate scholars, who defend the concept that humans evolved from other animals, should not be neglected. In fact, there are many young Muslim science students and professionals who are 'willing to accept evolution as a mechanism for the emergence of all species except humans' (Dajani, 2015). It is true that many of them are inclined to re-interpret Islamic texts so as to reconcile transmutation theories with their own religious beliefs (Asghar, 2013). But that is exactly what some well-known Western evolutionary biologists do when they try to combine Christian beliefs with biological evolution, or at least to defend it by saying that the two are not incompatible. This is the case, for instance, with Francisco Ayala, who discusses it extensively in his 2007 book, *Darwin's Gift to Science and Religion*.

With the points outlined above in mind, we will now provide a review of the evolutionary ideas of some key pre-Darwinian Muslim scholars. It is important to note that when we refer to 'evolutionary ideas', we are referring to the simplest definition of biological evolution being *the notion that species change over time*. That is, we are referring to the concept of transmutation, which was defended by many Western scholars, such as Lamarck, Erasmus Darwin, and so on, well before Wallace and Darwin defended the idea of evolution by natural selection. As we will explain, the transmutation ideas of some Muslim scholars were similar to those of some pre-Wallace and pre-Darwin Western scholars, while others were strikingly similar to a number of ideas defended by Wallace and Darwin concerning *natural selection*. It is also important to note that the texts discussed below are not *all* the existing sources referring to evolutionary ideas amongst pre-Darwinian Muslim scholars. Rather, we are focusing on the transmutation ideas of the *eight key Muslim scholars* listed in **Fig. 3.2** and shown in **Fig. 3.3**.

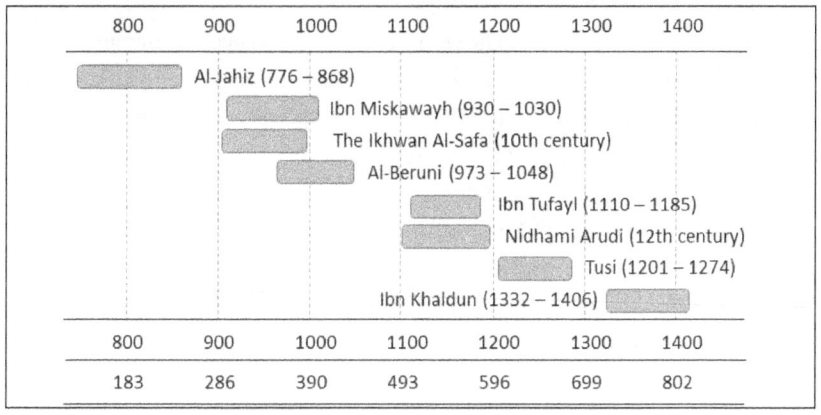

Fig. 3.2. Timeline of key pre-Darwinian Muslim scholars with evolutionary ideas from the 8th to the 13th centuries. The lower row shows the years in *Hijri*, that is, according to the Islamic calendar.

Fig. 3.3. Key pre-Darwinian Muslim scholars that wrote evolutionary ideas: **A)** Al-Jahiz (776–868); **B)** Ibn Miskawayh (930–1030); **C)** Ikhwan Al-Safa (10th century); **D)** Al-Beruni (973–1048); **E)** Ibn Tufayl (1110–1185); **F)** Nidhami Arudi (12th century); **G)** Tusi (1201–1274); **H)** Ibn Khaldun (1332–1406).

These scholars are those who, according to our overall literature search, had more articulated and/or developed ideas. That is, they produced more written texts and explicit statements about transmutation. Obviously, texts of other Muslim authors, such as the great poet, Rumi **(Fig. 3.4)**, are also interpreted by some current scholars as conveying ideas that are similar to those of the eight key Muslim scholars. For instance, Rumi wrote: "I died as a mineral and became a plant, I died as a plant and rose to animal, I died as an animal and I was Man. Why should I fear? When was I less by dying?" If this sentence had been written by a Western philosopher or biologist in the last one hundred and fifty years or so, one could say it brilliantly puts into perspective death and life, and how things, including species, change through the cycles of death and reproduction in the natural world. Of course, we are not saying that Rumi was, literally, trying to say this, but what he wrote is true in the face of what we know about biological evolution today, except that minerals did not evolve into plants, obviously (although inanimate particles and chemicals did combine to form life), nor did plants into animals (they are two different kingdoms within the natural world), but within the old notion of the ladder-of-life, which was how most thinkers saw the ladder **(Fig. 1.9)**, as we will see below.

Fig. 3.4. According to some interpretations, Rumi, the great 13th century poet, shown in this Ottoman era manuscript meeting Shams-e Tabrizi, referred to different stages of humans, such as a first stage from inanimate matter and a second stage from other animals.

On the ladder-of-life, each group or kingdom leads to the next, as the ladder is seen as being linear, while in the reality of the natural world, biological evolution is often more like a tree, with different branches in which any two groups can be sister-groups. For instance, living organisms, humans and chimpanzees are sister-groups, because they are derived from a common ancestor. Thus, humans were not derived from chimpanzees, nor were chimpanzees derived from humans. However, because many Muslim scholars were highly influenced by Aristotle's writing about the natural world, a recurrent theme in their writings was the notion of such a linear ladder-of-life, as described by Ibn Miskawayh, Al-Beruni, the Ikhwan Al-Safa, Nidhami Arudi, Tusi, and Ibn Khaldun. These scholars proposed a ladder-of-life hierarchy consisting of minerals, plants, animals and humans, with each 'kingdom' giving rise to and/or serving the next (Anderson, 2004; Diamandopoulos and Goudas, 2007; Ibn Khaldun, 1377; Nasr, 1993; Nidhami-i-Arudi-i-Samarqandi, 12th century). However, Aristotle and other Greek scholars, as noted in **Chapter 1**, often referred to the ladder-of-life as purely hierarchical, as if the natural world had been created exactly as it was when they wrote their texts, the hierarchy being part of a master plan of the Gods that made it, or of Mother Nature, or another type of creator. Aristotle was ambiguous about this, with some of his writings suggesting that there was actually transmutation between some forms of life, as noted in Leroi's superb book, *The Lagoon – How Aristotle Invented Science*. But none of his texts stated this in a consistent, clear way. Some of the texts of the key Muslim scholars discussed below are also ambiguous about these topics, as we will see. However, other texts seem to be much clearer about, not only the occurrence of transmutation, but even its causes, including some causes that resemble certain ideas of Wallace and Darwin about natural selection, as noted above.

Al-Jahiz

Al-Jahiz, born Abu Uthman Amr Bin Bahr Al-Fukaymi Al-Basri (776–868) in Iraq, was an 8th century Muslim zoologist (Bayrakdar, 1983; Guessoum, 2011). Learned in subjects including philosophy, literature, and Arabic, he earned recognition for his *Kitab Al-Hayawan* (*Book of Animals*, **Fig. 3.5**) (Bayrakdar, 1983). This book detailed his evolutionary theory in relation to the natural selection of animals, particularly birds (Guessoum, 2011). Al-Jahiz is widely considered to be the first Muslim scholar to propose an evolutionary theory which, according to some sources, was also the first in the field of science (Bayrakdar, 1983). Al-Jahiz devoted years to the scientific study of animals, closely observing their behaviour and classifying them by their characteristics and similarities (Bayrakdar, 1983).

MUSLIM SCHOLARS AND EVOLUTIONARY IDEAS 125

Fig. 3.5. A giraffe from Al-Jahiz's *Kitab al-Hayawan* (*Book of the Animals*).

Al-Jahiz's theories are said to have impacted the fields of zoology and biology, and heavily influenced Arab Muslim and European thinkers alike (Bayrakdar, 1983). Later scholars, such as Ibn Miskawayh and Ibn Khaldun, are thought to have based their own theories on his ideas. Our comparative research supports the notion, also proposed by other researchers (e.g., Bayrakdar, 1983), that there are obvious parallels between Al-Jahiz's ideas and some of those of Wallace and Darwin regarding natural selection. Some authors go as far as to say that Darwin and other 18th and 19th century scientists used Al-Jahiz's work as a foundation for their own evolutionary theories, albeit in a more modern scientific format (Bayrakdar, 1983). However, this can only be discussed in detail when historians find a direct link between Al-Jahiz's writings that Wallace and Darwin may have read, or the writings of Western authors that they read. So far, according to our knowledge, despite the fact that some authors argue that they have found a direct link, this is far from being consensual among Western and even among non-Western historians. Thus, more studies around this issue are clearly needed. However, it should also be noted that those who argue that Al-Jahiz's transmutations were 'inconsequential' because they seemingly did not 'reach' and thus influence Western scholars such as Wallace and Darwin, are seeing things from a Western perspective. It is as if observations are only worthy if they 'reach' or are 'useful' to Westerners, the pinnacle of evolution. If the first spaceship to take humans to Mars was to be completely made and operated by Chinese people, this would not be less remarkable or historically relevant just because it did not involve Westerners. Similarly, Sputnik 1, the first artificial Earth satellite, was no less significant or outstanding because it was launched by the Soviet Union, and not the U.S., U.K., Germany or France.

So, what did Al-Jahiz write about this topic? Regarding transmutation, he explained the great impact of environmental conditions on the ability of organisms to adapt, i.e., they develop new traits that subsequently allow them to survive in those new conditions. However, an important difference between his ideas and those of Wallace and Darwin is that Al-Jahiz took God's role into account, believing that organisms ultimately evolve by Divine will and guidance, developing new traits for survival as part of God's plan to keep nature in order. In contrast to Wallace and Darwin, who did not refer to biological evolution as a master plan of God—although Darwin often referred to natural selection as Mother Nature's Broom—many other evolutionary biologists did, and do, refer to evolution in such a way. In this sense, Al-Jahiz's view is not so different from Lamarck's, who, in his book *Philosophie Zoologique*, argued that God was the 'sublime author of nature'.

Al-Jahiz described a natural selection-like process resulting from an animal's innate desire to live, stating that biological fitness is essential to this phenomenon (Saniotis, 2012). He observed that individuals of the same species compete against each other and that the stronger, better adapted ones will prevail as they are likely to have lower mortality rates. His understanding of the circle of life is shown in an example of a rat that feeds on smaller animals and digs underground dwellings for protection against predators. Those predators, in turn, have their own defences against even larger predators, and so forth. Al-Jahiz wrote: 'This is the law that some existences are the food for others... all small animals eat smaller ones; and all big animals cannot eat bigger ones. Men with each other are like animals' (Bayrakdar, 1983). Zirkle (1941) further discusses this topic using this same passage from Al-Jahiz, translated by the Spanish writer, Miquel Asin Palacios: 'All animals, in short, cannot exist without food... Every weak animal devours those weaker than itself. Strong animals cannot escape being devoured by other animals stronger than they. And in this respect, men do not differ from animals, some with respect to others, although they do not arrive at the same extremes. In short, God has disposed some human beings as a cause of life for others, and likewise, he has disposed the latter as a cause of the death of the former.'

Al-Jahiz recognised the role of environmental conditions in determining human skin colour, after noting the skin tone variations between northern and southern peoples (Kechichian, 2012). Furthermore, he wondered about the evolution of *al-miskh*, or the original quadrupedal (tetrapod) ancestor: 'People said different things about the existence of *al-miskh*... some accepted its evolution and that it gave existence to dog, wolf, fox and their similars. The members of this family came from this form' (Bayrakdar, 1983). Another portion of Al-Jahiz's texts alludes to human evolution and the similarities between animals and apes: 'We have seen that some *nabatheen* [Nabataeans] navigators resembled the ape in some geographical environment; likely we have also seen some people from Morocco and have found them as like *al-maskh* [a kind of ape, not to be confused with *al-miskh*], except for a little difference... And it is possible that the polluted air and water, and dust made this change in the character of these Moroccans... those changes in their bristles, ears, colours, and form (similar to apes) increase more...' (Bayrakdar, 1983).

Ibn Miskawayh

Ibn Miskawayh (930–1030) was a 10th century Persian Muslim philosopher whose evolutionary ideas may have been influenced by those of Al-Jahiz (Bayrakdar, 1983; Hawi, 1974). Hawi cites a passage from Ibn Miskawayh's *Al-Fawz Al-Asghar*, stating that Ibn Miskawayh 'possessed a profound awareness of the evolution of life that stands on a par with the views of Darwin, Huxley and others.' In that passage, Ibn Miskawayh wrote:

> The first step of the ascension of plants, of a higher order, is to free themselves from the ground and from their need to consolidate their veins in it... This first animal stage is weak... Animals in this stage remain weak in locomotion even though they have freed themselves from the ground and evolved a new life... Then they evolve from this to another stage: here their capacities of movement and sense become stronger; such is the case of worms, many kinds of butterflies and crawling beings... Sensitivities in these new animals becomes stronger and from them emerge animals having four senses such as the mole and the like... From here they [the animals] progressively evolve to a higher stage in which sight is generated; this is the case with ants and bees... then they approach the last stage of the animal kingdom. Although this rank is superior, nevertheless it remains base and inferior, far from the level of monkeys and the like. These become near to man in structure and human appearance. There is no difference between those types and man except a little, which if surpassed, they become humans.

However, after reading this excerpt, we can see that Hawi's assertion that there are similarities between Ibn Miskawayh and Darwin and Huxley, is flawed. Firstly, Darwin never stated that animals came from plants, nor that we are derived from ants and bees. Instead, he said that we share a common ancestor with all those organisms. In a simplistic way, humans are like the cousins of ants and bees, and not their sons or grandsons. Secondly, Hawi, as with many non-Western scholars, refers to Darwin and his 'bulldog', Huxley, but not to Wallace. Evolutionary thought is much more than Darwin and his 'bulldog', Huxley. One cannot criticise Western scholars for following a simplistic narrative while neglecting the writings of certain key Muslim scholars, as well as Wallace, who, incidentally, was the first to finish a full manuscript about evolution by natural selection (in 1858, before Darwin's 1859 *On the Origin of Species*).

Where Ibn Miskawayh's excerpt is similar to what Wallace, Darwin, Huxley and other Western scholars wrote, is his suggestion that humans evolved from other animals, the main difference between them being the intellect which God gifted to humans (Kaya, 2012). He declared that changes in the human lineage brought about 'psychological changes with the growing power of discrimination and spirituality until humankind passed from barbarism to civilization' (Shanavas, 2010). Hawi's belief that Ibn Miskawayh possessed an understanding of biological evolution similar to these Western scholars was shared by Muhammad Iqbal, a renowned Indian scholar and poet of the 20th century (Hameed, 2008).

According to Shoja (pers. com.), Ibn Miskawayh also discussed other aspects of transmutation, such as: '[These books] state that God first created matter and invested it with energy for development. Matter, therefore, adapted the form of vapor which assumed the shape of water in due time. The next stage of development was mineral life. Different kinds of stones developed in the course of time, their highest form being marjan [coral]... After mineral life evolves vegetation. The evolution of vegetation culminates with a tree which bears the qualities of an animal... Then is born the lowest of the animals. It evolves into an ape... That ape then evolved into a lower kind of barbarian man... He then became a superior human being.'

The Ikhwan Al-Safa

The Ikhwan Al-Safa (the Brethren of Purity) was a 10th century esoteric society of Arab Muslim thinkers based in Basra, Iraq (Guessoum, 2011; Hameed, 2012). While they were deemed heretical by orthodox communities in Iraq, the Ikhwan's work 'spread as far as Spain, where it influenced philosophical and scientific thought' (Singer 1941). Often referred to as a Platonic, or Neoplatonic, Shi'ite, 'Pre-Sufi' group of scholars (Guessoum, 2011; Siddiqi, 1995), the Ikhwan are known for their work *Rasa'il Ikhwan al-Safa* or *Epistles of the Brethren of Purity*, in which they explored subjects including cosmology, philosophy, logic, theology and mathematics (Săvoiu, 2014).

Nasr (1993, p.70) notes that the Ikhwan grouped all matter into three 'kingdoms' (minerals, plants, and animals), with members of each one being 'connected to the first member of the next domain... Minerals are connected below to water and earth, and their lowest types are alum, hyacinth, and vitriol, which are very close to earth. Red gold, on the other hand, stands highest among the minerals and approaches the world of the plants... moss is the lowest order approaching the mineral kingdom, while the palm tree, which already has a differentiation of sexes, stands between the plant and animal worlds.' These ideas are comparable to the *scala naturae*, which is the great chain of being or the ladder-of-life concept (**Fig. 1.9**) developed by Greek philosophers such as Aristotle.

The Ikhwan described animals in relation to humans as follows: 'Among animals, the snail is mentioned as being closest to the plant world and the elephant being highest in intelligence among the animals, nearest to man' (Nasr, 1993). Nasr explains that: 'Inasmuch as this hierarchy is based on the degree of intelligence and the development of internal faculties rather than on external similarities, we find that the Ikhwan name the elephant rather than the monkey as the closest animal to man… this is a good example of the difference between the traditional idea of gradation, which is based on internal qualities and ontological status, and the modern theories of evolution which are based on the physical behaviour and the external similarities of creatures.' In fact, scientists often study apes and other animals, in order to infer the evolution of the human brain, because apes are our closest living relatives. However, it is consensually accepted that elephants are mentally highly complex, being in some respects remarkably similar to humans, and in other respects unparalleled throughout the animal kingdom (Garstang, 2015).

The Ikhwan considered life to have developed gradually, and that minerals, plants and animals existed before humans did (Shanavas, 2010). They emphasised that God created animals for the benefit of humans, and that He equipped animals with all the necessary means to live. According to Shanavas, they said that:

> The first stage of the plant kingdom is connected with the last stage of minerals and the highest stage of the plant kingdom with the first stage of animal… the highest stage of the animal kingdom is connected with the first stage of the human. Be it known to you! The imperfect animals preceded the most perfect animals *in time* and in the process of creation… The animals were created for man's sake. [And] everything that is created for the sake of something else will precede the beneficiary. By the grace of God's wisdom and care, animals were bestowed with organs, joints, vessels, nerves, membranes, and chambers according to their needs for benefitting themselves or to avoid injury, so that it can survive and will be completed and perfected to reach the highest stage.

Although some have claimed that the Ikhwan referred to biological evolution, Nasr (1993) states that their 'divergence from modern theories of evolution should be clear.' They asserted that all changes occurring on Earth are caused by the 'Universal Soul' (God), and that one species cannot change into another: 'The species and genus are definite and preserved. Their forms are in matter. But the individuals are in perpetual flow; they are neither definite nor preserved. The reason for the conservation of forms, genus, and species, in matter is the fixity of their celestial cause because their efficient cause is the Universal Soul of the spheres instead of the change and continuous flux of individuals which is due to the variability of their cause' (Nasr, 1993). Furthermore, the Ikhwan were adamant in their belief that creation occurred both sequentially and instantly, in an 'emanation' process from God (Guessoum, 2011).

Nevertheless, there are crucial differences between what we now know about biological evolution and what the Ikhwan said. For instance, they suggested that 'the date of the beginning of terrestrial existence of plants precedes that of animals, just as minerals precede the plant.' Of course, we now know that plants evolved about 500 million years ago, while animals evolved about 800 million years ago. This shows how wrong the teleological concept of the ladder-of-life is when applied to biological evolution. The Ikhwan also wrote: 'Plants come before (*taqaddama*) animals in the series of beings and serve them as material for the forms of animals and food for the nutrition of their bodies... plants would be like a mother who eats raw food... and transforms it into pure milk which is absorbed very gently by those who drink it. The plants subsequently present this to the animals considered as their sons... Plants occupy an intermediate position—necessary and salutary—between the four elements and the animals... All the parts of the vegetables which the animals consume such as seeds, leaves, fruit and so on, come from the four elements digested and transformed by the plants' (Nasr, 1993). This is indeed a teleological idea that views the world as a masterplan in which some organisms are made to 'service others': minerals preceded—this is correct—and serviced plants, while plants preceded—this is incorrect—and serviced animals, who, in turn, service humans (who, in reality, are animals): 'who therefore comes to this world later than all of them, since each has come after the kingdom on which it depends.' This latter part is also only partially correct, because, while it is true that humans are relatively recent evolutionarily, as we and chimpanzees split about 7 million years ago, other groups of animals are even more recent than we are, i.e., we are *not* at the 'pinnacle' of evolution. This also applies to our species, *Homo sapiens*, which evolved about 300 thousand years ago, and, thus, is older than many other species, such as dogs who evolved much more recently, and those that are evolving right now. Evolution is dynamic, not written in stone, and humans are not the last ones, or the best ones, or the 'chosen' ones, or the most perfect ones; we are just one species of the millions of species that have existed in this wonderful planet.

Al-Beruni

Abu Al-Raihan Muhammad Ibn Ahmad Al-Beruni (973–1048) was an 11th century Muslim astronomer, mathematician, geographer, and historian from what is now Turkmenistan and Uzbekistan (Singer, 1941; Starr, 2009). He lived in India and aspired to 'control the Hindu language in order to impart its knowledge to Arab-speaking nations at the beginning of the eleventh century' (Wilczynski, 1959). One of his most famous works, entitled *India* (Wilczynski, 1959), gained him recognition as the first Muslim and one of the most prominent scholars to objectively write about Hinduism (Guessoum, 2011; Sharma, 1991). An 'independent thinker' who viewed the study of nature as a religious duty, he created theories that adhered to both Quranic teachings and a scientific approach (Guessoum, 2011).

Wilczynski commented on Al-Beruni's theories, saying: 'Nature performs natural selection of the most adequate, well-adapted beings through the extermination of others, and, in this case, it proceeds in the same way as farmers and gardeners... Darwin's great idea of natural selection through the struggle for life and survival of the fittest was already reached by Al-Beruni approximately eight hundred years before Darwin. It is true that he seized it in the most general outlines only, but, curiously enough, even the very meaning and the way in which he came to it were the same as Darwin's' (Rainow, 1943). Again, we do not think such comparisons are always useful as they not only equate evolutionary biology with Darwin, but also seem to imply that Muslim scholars could only have been great when they had similar ideas to Western scholars. The important point we want to make in this book is that some key Muslim scholars had, during the Golden Islamic Age, transmutation ideas.

Al-Beruni expressed the idea that man 'migrated' through the 'kingdoms' of minerals, plants, and animals 'in order to reach perfection and therefore contains within himself the nature of the creatures of the other realms' (Nasr, 1993). According to Nasr, his work *Kitab al-Jamahir*, hints at evolutionary ideas, namely that monkeys were the final stage from which mankind 'migrated' to reach his present state. Al-Beruni wrote: 'Man reached his maximum degree of perfection compared to other animals below him... He ascended to the present state from other kinds of beings such as dog-like, bear-like, ape-like, etc. Then finally he became man' (Shanavas, 2010). While Al-Beruni was clearly influenced by the Ancient Greek concept of the ladder-of-life, and may have been influenced in other ways by Hellenistic thinkers (Hameed, 2012), his evolutionary ideas are present in his book about Hinduism and the 'general natural processes in the whole world' in a chapter entitled, *On Vasudeva and the Wars of the Bharata* (Wilczynski, 1959). His transmutation views are summarised in passages about 'four different phenomena', the first of which is: 'The life of the world depends upon the sowing and procreating. Both processes increase in the course of time, and this increase is unlimited, whilst the world is limited' (Wilczynski, 1959). The second passage addresses Al-Beruni's ideas on species expansion: 'When a class of plants or animals does not increase any more in its structure, and its peculiar kind is established as a species of its own, when each individual of it does not simply come into existence once and perish, but besides procreates a being like itself or several together, and not only once but several times, then this will as a single species of plants or animals occupy the earth and spread itself and its kind over as much territory as it can find' (Wilczynski, 1959). This passage suggests that Al-Beruni was thinking about biological reproduction, speciation, and some kind of differential of survival, eight hundred years before Wallace and Darwin (Shanavas, 2010).

A third excerpt refers to artificial selection and social animals: 'The agriculturist selects his corn, letting grow as much as he requires, and tearing out the remainder. The forester leaves those branches which he perceives to be excellent, whilst he cuts away all others. The bees kill those of their kind who only eat, but do not work in their beehive' (Wilczynski, 1959). Wilczynski suggests that the fourth relevant passage is comparable to Darwin's theory of natural selection, because Al-Beruni wrote: 'Nature proceeds in a similar way; however, it does not distinguish for its action is under all circumstances one and the same. It allows the leaves and fruit of the trees to perish... It removes them so as to make room for others.' In this case, there is, indeed, a striking similarity with Darwin's writings about natural selection being the 'broom' of Mother Nature—more so than with Wallace's publications. Wallace's view of nature and natural selection was not as markedly teleological as Darwin's, for example, when the latter wrote in *On the Origin of the Species*, that Mother Nature 'is daily and hourly scrutinizing, throughout the world, every variation, even the slightest; rejecting that which is bad, preserving and adding up all that is good.'

Of course, there is a major difference: Darwin's Mother Nature was more part of a teleological narrative, while Al-Beruni attributed *all* natural processes, as well as any 'imperfections' that occur during an animal's evolution and development, to God's infinite wisdom. He wrote: 'When Nature does not find the substance by which to complete the form of that animal in conformity with the structure of the species to which it belongs... she forms the animal in such a shape, so that the defect is made to lose its obnoxious character, and she gives it vital power as much as possible... Frequently... you find in the functions [actions] of Nature which it is her office to fulfil, some fault [some irregularity], but this only serves to show that the Creator who had designed something deviating from the general tenor of things is indefinitely sublime, beyond everything which we poor sinners may conceive and predicate Him' (Shanavas, 2010). This argument, also often used by Christian Creationists, is recurrently used by many Muslim scholars who tend to attribute a more active role to God concerning what is happening to our planet. According to this argument, if something goes 'wrong', either in evolution or in development—for instance, a young child having a congenital malformation or cancer—it is probably part of God's masterplan, to test the faith of the parents, or the whole family, or of humanity, in general.

However, we need to emphasise that, as discussed above, the comparisons between the importance that God assumed within the transmutation ideas of Muslim authors, such as Al-Beruni, versus the relevance that God or Mother Nature assumes within the transmutation ideas of Western scholars, such as Darwin, should take into account that the former lived centuries earlier than the latter. If the former were compared, instead, with Western scholars or philosophers of the 10th century—as Al-Beruni was—then there is no doubt that the latter were, in general, more obsessed with God than scholars such as Al-Beruni, although there are obviously exceptions, as noted above.

Ibn Tufayl

Abu Bakr ibn Tufail, also known as Abubacer or Ibn Tufayl (1110–1185), was an Andalusian Muslim scholar born in Gaudix, Spain, during the early 12th century (Cerdá-Olmedo, 2008). A physician and teacher of the Andalusian philosopher, Ibn Rushd (Averroes), Ibn Tufayl is most famous for his Arabic philosophical novel entitled, *Risala Hayy ibn Yaqzan fi asrar al-hikmat al-masriqiyya*, or simply, *Hayy ibn Yaqzan*, which translates as: *Philosophus Autodidactus, The Story of Alive, Son of Awaken, On the Secrets of Oriental Wisdom*, or simply, *The Journey of the Soul* in Latin and English, respectively (Cerdá-Olmedo, 2008).

Widely recognised for its influence on Arabic and Islamic philosophy and literature, as well as Western culture between the 16th and 19th centuries, this book is thought to have inspired Daniel Defoe's *Robinson Crusoe* (Attar, 2007), Rudyard Kipling's *The Jungle Book,* and other works about human beings living in isolation (Cerdá-Olmedo, 2008). The first Latin and English translations are thought to have been published in 1671 and 1708, respectively, followed by those in German, Dutch and French (Shanavas, 2010). It tells the story of Hayy, a boy growing up on an island with only animals for company as he struggles to survive, defends himself against nature, and discovers philosophical truths (Hawi, 1974). The tale's themes are reportedly based on Ibn Tufayl's own philosophical beliefs, suggesting that he believed in a naturalistic origin of life but also hinting at evolutionary ideas (Hawi, 1974). It is unclear what specific evolutionary ideas the scholar subscribed to but, according to Hawi, Ibn Tufayl's writings show that he believed 'in the generation of life from matter and that man is a product of the interworking of the natural elements, whether man was generated spontaneously or was a result of an evolutionary process.'

Ibn Tufayl described the boy, Hayy, as spontaneously originating from a hot, damp, 'fermented' clay 'that labored and churned like bubbles over boiling water' over the course of several years. This clay had the 'suitability to be formed into all the protective membranes and the like which would be needed in the forming of a man. When the embryo was complete these coverings were sloughed off as if in labor; and the clay which had already begun to dry, cracked up' (Hawi, 1974). According to Hawi, these descriptions suggest that Ibn Tufayl was writing about a type of biogenesis: that organic compounds existed before life in an 'inorganic world', and that inorganic compounds slowly transform, leading to 'a spasmodic emergence of life'.

It should be noted that this idea of change from an inorganic to organic world is expressed in both the Bible (Genesis 2:7) and several Quranic verses, which state that all life forms were created from water and that human beings were shaped from clay (Quran, 21:30, 24:45, 25:54, 55:14; Asghar *et al.* 2010). What is more peculiar about Ibn Tufayl's story is that his descriptions of clay and boiling water are similar to one of modern science's most discussed hypotheses about the origin of life: that is, that life originated in hot water over a long period of time. In addition, by describing the gradual morphing of clay, he also seems to have thought, as modern scientists do, about 'the very lengthy period that the... generation of living beings takes... Life does not arise spontaneously overnight' (Hawi, 1974). Furthermore, according to Hawi, Ibn Tufayl theorised that inanimate matter is the foundation for all forms of life and that balance is essential to life. In this sense, it is true that the concept of long periods of geological time is in stark contrast to ideas now defended by most creationist religious fundamentalists, who incline towards the view that all life was created almost instantly within a geological timeframe.

The details of Hayy's origins strongly capture Ibn Tufayl's scientific knowledge. Hawi wrote this about Ibn Tufayl's descriptions: 'Under the influence of heat and sunshine, the embryo is formed. These were not wild guesses, but facts dictated by the searching and observant mind of a scientist. The fermented clay, viscous mass, light and heat are hypotheses presented by Ibn Tufayl to account for the emergence of life; but so are the modern interpretations by Darwin, Haldane and Keosian in terms of "warm ponds", ammonia, phosphoric salts, light, heat and electricity... Ibn Tufayl is equally aware as these scientists are of the effect of light and heat in enhancing the preparatory chemical process.'

Furthermore, it is important to note—particularly in the context of assertions made by many current Western scholars that the scientific method only truly arose in Europe a few decades ago (see **Chapter 1**)—that Hayy's use of observation and reasoning throughout his journey resembles current scientific thinking. Cerdá-Olmedo (2008) wrote about Ibn Tufayl's story: 'The idea that knowledge must be acquired through observation, experimentation, and thought was not new, but it had not been expressed as clearly and forcefully before.' Hayy's dissections of animal cadavers, in order to try to understand the 'principle of life', may also have stemmed from Ibn Tufayl's knowledge of anatomical dissections (Cerdá-Olmedo, 2008).

According to Hawi, in describing Hayy's attempts to compete for food and protect himself, Ibn Tufayl evidently understood adaptation to one's environment and the struggle for existence, which are evolutionary concepts that modern textbooks attribute to scholars such as Wallace and Darwin. Upon seeing that animals possess horns, claws, and other features, Hayy contemplates his own lack of natural defences but soon discovers the means to fashion his own weapons and tools (i.e. spears, shields), and begins training animals to service him. Ibn Tufayl wrote: 'This was due to [Hayy's] realisation that despite his lack of natural weapons, he could manufacture everything to make up this lack... There were wild horses on the island as well as wild asses... Then out of thongs and rawhide he contrived saddles and bridles. So, as he had hoped, he was able to chase animals he had found difficult to catch' (Hawi, 1974).

However, some assertions about Ibn Tufayl seem far stretched. For instance, nothing mentioned in the above paragraph refers to evolutionary change per se: all those things would equally apply in a world without transmutation. Moreover, even those ideas of Ibn Tufayl's that imply evolutionary change, such as the origin of humans from inanimate objects, appear to resemble spontaneous generation, and do not explicitly define him as an evolutionist. This was recognised even by Hawi, who pointed out: 'To pronounce him an evolutionist, in the full Darwinian sense, is to place him in a naturalistic perspective for which his brief remarks do not qualify him.' Nevertheless, Ibn Tufayl's commentary on the origins of life as *a whole*—i.e., *not* the origins of humans, in particular—from inorganic matter seems to indicate an evolutionary view; that is, that life was not static or hurriedly made in a few days by a God (Hawi, 1974).

Nidhami Arudi

Ahmad bin Umar bin Ali an-Nidhami as-Samarqandi, Nidhami-i-Arudi-i-Samarqandi, or Nidhami Arudi of Samarqand, was a 12th century Persian Muslim thinker and poet who explored the natural sciences, philosophy and politics in his book entitled, *Chahar Maqala* (*Four Discourses*) (Khodadoust et al. 2013). Browne's (1999) English translation of *Chahar Maqala* was used to study Arudi's ideas of evolution (Singer, 1941). On the origins of vegetation, Arudi wrote: 'But when time began... and the composition of this lower world became matured, and the time was come for the fertilization of that interspace which lay between the water and the air, the vegetable world was manifested. Then God, blessed and exalted is He, created for that substance whereby the plants were made manifest four subservient forces and three faculties' (Nidhami-i-Arudi-i-Samarqandi 12th century, p.14). Based on our own understanding of this text, these four 'subservient forces' appear to represent Arudi's theories of organismal function, while the 'three faculties' refer to how organisms sustain themselves and reproduce.

The first two 'faculties' involve nutrition (i.e., how a plant attains and distributes nutrients to each of its structures), while the third refers to reproduction. Arudi wrote: 'When the organism has attained perfection and begins to tend towards defect, appears and produces germs, in order that, if destruction overtake the parent in this world, these may become its substitute and representative, so that the order of the world may be guarded from detriment, and the species may not cease. This is called the *Reproductive Faculty*.' That is, Arudi is referring to how organisms function, either by ensuring the survival of their species or adapting to become other new species, thereby contributing to a certain balance within the natural world.

Like the Ikhwan Al-Safa two centuries before him, Arudi discussed God's creation of three 'kingdoms': 1) minerals, 2) plants and vegetables, and 3) animals. He alluded to animals' superiority over minerals and plants, theorising that coral was the first organism to undergo a transformation from mineral to living plant: 'So this kingdom rose superior to the mineral and vegetable kingdoms in several ways... and the far-reaching Wisdom of the Creator so ordained, that these kingdoms should be connected successively and continuously, so that in the mineral kingdom the first thing which attained completeness and underwent the process of evolution became higher in organisation until it grew to coral... which is the ultimate term of the mineral world, until it was connected with the first stage of plant life.'

Also, like the Ikhwan, Arudi stated that date palm trees are the most superior members of the plant kingdom due to their methods of reproduction. He explained: 'And the first thing in the vegetable kingdom is the thorn, and the last the date palm, which has been assimilated to the animal kingdom since it needs the male to fertilise it so that it may bear fruit.' Then, once the plant kingdom had 'attained perfection', a 'finer offspring resulted, and the manifestation of the animal world took place'. The animal kingdom not only retained all the 'faculties' of the vegetable and plant kingdoms, but also gained two more: the 'Perspective Faculty', which is further divided into the five senses (touch, taste, smell, hearing and sight), memory, imagination, and cognition, and the 'Motor Faculty', which allows animals the power of voluntary movement. Again, Arudi is following the Greek notion of a ladder-of-life. The difference between him and many Greek scholars is that he is less ambiguous about this ladder being part of a masterplan of a God or Mother Nature. Instead, he clearly implies that there *was* biological evolution on this planet which mirrored this masterplan. But, of course, once again, despite postulating organic evolution with ancestors and their progeny in a much clearer manner, the specifics of his writings do not fit modern evolutionary knowledge because animals did not evolve from plants, as we have seen above.

Within this transmutational ladder-of-life, Arudi went on to describe a hierarchy within the animal kingdom, stating that 'perfect' animals possess all the Perspective and Motor Faculties, and that ants, snakes, maggots and worms are the lowest, 'defective' members of the hierarchy because they lack sight and hearing. He also mentioned a creature called a satyr (*nasnas*), being of 'erect carriage, of vertical structure, with wide flat nails', an intermediate between animals and humans, and the highest-ranking member of the animal kingdom after mankind. Older literature often uses the word 'satyr' to refer to apes; for instance, orangutans have been scientifically designated as '*Simia satyr*' (Owen, 1830). Some current authors suggest that the Ancient Greeks and Muslims already knew about African apes—chimpanzees and gorillas—in Arudi's time. If this was so, then Arudi's idea would be notable as it suggests a direct connection between apes and humans. However, 'satyr' was also a term commonly used in the past to designate a fictional half-human and half-'beast' being, so one cannot be sure if Arudi was referring to apes or to such fictional beings. Studies are needed to uncover what the Ancient Greeks, Romans, and Muslims during the Islamic Golden Age knew, or did not know, about apes.

Tusi

Nasir ad-Din Tusi (1201–1274) was a 13th century Persian Muslim thinker who, according to authors such as Saniotis, defended some aspects of evolution. While known for his Arabic and Persian writings on medicine, philosophy, geography, and ethics, Tusi's theories on evolution are presented in his work *Akhlag Nasiri* (Nasirean Ethics). In it, he discusses his belief in mankind's spiritual and material perfection (Alakbarli, 2001). His works use the word *takamul* recurrently, which is translated to mean 'perfection' or 'evolution' in Arabic and Azeri (Alakbarli, 2001). As with many other Muslim scholars discussed in the present book, Alakbarli states that Tusi was probably not credited as an evolutionist by Western scientists and historians as they tend to interpret his ideas as being based upon Islamic philosophy, thereby assuming his theories are more religious than scientific or naturalistic.

Also, according to Alakbarli, the evolutionary ideas of ancient Greek scholars, such as Empedocles and Aristotle, were further built upon by Muslim writers before Tusi, including Al-Beruni and Ibn Tufayl (see above). Tusi, in turn, was influenced by these scholars and, according to some researchers, 'foreshadowed' the theories of Western evolutionary biologists with his approach (Alakbarli, 2001). Alakbarli wrote that he moved 'from theory to facts, instead of the other way around... When he wrote about evolution (he called it "perfection") as a theory, he therefore didn't dwell on the details... For instance, he didn't write specifically about natural selection or the struggle for existence. In modern terms, he was more of a philosopher than a scientist.'

Tusi's theories on hereditary variability (Saniotis, 2012) and an organism's ability to change according to environmental conditions have some vague similarities with current concepts on the importance of both genetic and epigenetic factors within natural selection. Accordingly, he wrote: 'The organisms that can gain the new features faster are more variable... As a result, they gain advantages over other creatures... The bodies are changing as a result of the internal and external interactions.' Tusi further observed that such changes allowed organisms to survive: 'Look at the world of animals and birds. They have all that is necessary for defence, protection, and daily life, including strengths, courage and appropriate tools [organs].' Additionally, Tusi described a hierarchy of living things in nature, following the older concept of the ladder-of-life: 'Animals are higher than plants, because they are able to move consciously, go after food, find and eat useful things... There are many differences between the animal and plant species... the animal kingdom is more complicated. Besides, reason is the most beneficial feature of animals. Owing to reason, they can learn new things and adopt new, non-inherent abilities. For example, the trained horse or hunting falcon... is at a higher point of development in the animal world. The first step to human perfection begins here.'

Tusi wrote about transitions between humans and animals: 'Such humans live in the Western Sudan and other distant corners of the world. They are close to animals by their habits, deeds, and behaviour.' This idea was endorsed by Diamandopoulos and Goudas (2007), who pointed out that Tusi, not only echoed his predecessors' theory on the hierarchy of 'kingdoms', but that he also emphasised the similarities of humans and other animals, suggesting that there is a unity between all organisms: 'The human has features that distinguish him from other creatures, but has other features that unite him with the animal world, the vegetable kingdom or even with the inanimate bodies.' Moreover, according to Alakbarli (2001), Tusi suggested that 'the human being is placed on the middle step of the evolutionary stairway. According to his inherent nature, the human is related to the lower beings, and only with the help of his will can he reach the higher development level.' The ese of the term 'evolutionary stairway' by Alakbarli, seems to be a little far-fetched; a 13th century scholar obviously would not use such a phrase, or anything similar. It is more likely that this expression refers to the notion of the ladder-of-life—thence, stairway—which to some Muslim scholars was a physical, transmutational ladder. So, although the concept might be similar, using the term 'evolutionary stairway' seems a bit odd.

Ibn Khaldun

Ibn Khaldun (1332–1406) was a 14th century North African Muslim thinker, historian and polymath (Guessoum, 2011). Best known for his book *Al-Muqaddimah*, or *Prolegomena* in the West (Asghar et al. 2010), it contains ideas resembling modern evolutionary theory (Saniotis, 2012). Ibn Khaldun strongly emphasised human evolution. For example, he wrote:

> One should then look at the world of creation. It started out from the minerals and progressed, in an ingenious, gradual manner, to plants and animals. The last stage of minerals is connected with the first stage of plants, such as herbs and seedless plants. The last stage of plants, such as palms and vines, is connected with the first stage of animals, such as snails and shellfish which have only the power of touch... The animal world then widens, its species become numerous, and, in a gradual process of creation, it finally leads to man, who is able to think and to reflect. The higher stage of man is reached from the world of the monkeys, in which both sagacity and perception are found, but which has not reached the stage of actual reflection and thinking. At this point we come to the first stage of man after [the world of monkeys]. This is as far as our [physical] observation extends.

It is fascinating to see Ibn Khaldun writing that *'the higher stage of man is reached from the world of the monkeys'*, as this—whether said literally or symbolically—would be perceived as outrageous, or even heretical, by many Muslims nowadays, as well as those from other religions, including Christian creationists. On animal and human evolution, Ibn Khaldun further wrote: 'It is also the case with monkeys, creatures combining in themselves cleverness and perception, in their relation to man, the being who has the ability to think and to reflect... plants do not have the same fineness and power that animals have... Animals are the last and final stage of the three permutations. Minerals turn into plants, and plants into animals, but animals cannot turn into anything finer than themselves' (Ibn Khaldun, 1377).

Ibn Khaldun, thus, clearly seems to be defending, as do the other Muslim scholars reviewed above, the evolution of life over time. Shanavas's (2010) translation suggests that Ibn Khaldun stated that 'transmutations of one species into another over a long period of time resulted in the gradual evolution of life, from primitive organisms into a bush with numerous branches. Thus, life forms are not independently created but are evolutionary products from ancestral species.' Ibn Khaldun further rejected the Talmudic and Christian belief of dark African skin being a curse inflicted upon sinful human beings (Saniotis, 2012; Shanavas, 2010). Like Al-Jahiz in the 8th century, he observed the contrast in skin colour between northern and southern peoples, namely the dark skin of the Sudanese, and suggested a causal relationship between hot southern climates and dark pigmentation, an idea now known to be correct as the result of a combination of genetic and epigenetic factors (Saniotis, 2012). Regarding the hereditary changes that humans undergo, Ibn Khaldun wrote: '[There also is regard of the fact that physical circumstances and environment] are subjected to changes that affect later generations; they do not necessarily remain unchanged. This is how God proceeds with his servants... And verily, you will not be able to change God's ways' (Shanavas, 2010).

General remarks about the transmutation ideas of Muslim scholars

We have briefly summarised the transmutation ideas of eight of the most renowned Muslim scholars between the 8th and 14th centuries, showing that their theories were proposed in a *continuum* of time during the Islamic Golden Age. This was well before Western scholars such as Lamarck, Erasmus Darwin and, much later, Wallace and Charles Darwin, wrote about biological evolution. It is true that many of the current scholars we have referred to, particularly Muslim ones, overemphasise the 'similarities' between Western evolutionary biologists, particularly Darwin, and Muslim scholars of the Islamic Golden Age. This may be due in part to these current scholars having an over-simplistic knowledge of the history of Western evolutionary biology. For instance, any single reference to the ladder-of-life; a concept amongst scholars in Ancient Greece, is given as an example of an 'evolutionary idea'. However, it is clear that the eight Muslim scholars discussed in this chapter were less ambiguous about the ladder-of-life being the result of a real, physical change between the forms that developed according to this plan, than most Ancient Greek authors. In other words, the ideas of these Muslim scholars were derived from the *scalae naturae* concept of thinkers, such as Aristotle, but with one major difference. Unlike the atemporal (non-evolutionary) way in which the *scalae naturae* was interpreted by most Greek scholars, and subsequently by Christian and Muslim religious authorities, most of these Muslim scholars referred to *temporal* changes and different times of origin for different taxa and groups. That is, their theories were evolutionary because they referred to *the notion that species change over time*.

Moreover, these eight Muslim scholars suggested that humans underwent some type of phenotypic evolution. Some of them specifically wrote about similarities between humans and monkeys, in many cases suggesting that humans were derived from them. This is in complete contrast to the view defended nowadays by most Muslims and Christian creationists. However, it should be pointed out that all the Muslim scholars emphasised the belief that biological evolution and changes in nature occur by God's will and direct guidance. This is hugely different from the view held by many current evolutionary biologists, including those who are religious, many of whom purport that God is not directly involved in the specific changes occurring during biological evolution. In this sense, the ideas of those key Muslim scholars resemble those of current religious thinkers who defend the concept of *'intelligent design'*. Although, they recognise, in the face of the overwhelming empirical data available, that species change over time, they maintain that those changes were/are always determined/planned by a 'designer'.

Importantly, the fact that those Muslim scholars held such transmutation ideas remains unrecognised in both the Muslim world and modern Western science. This is, in part, due to bias; Eurocentrism and perceived conflicts between evolution and Islamic teachings, as explained above. Such views ignore the fact that those perceived conflicts originated mainly in the early 20th century before being manifested throughout the Muslim world in recent decades. Actually, some authors argue that the evolutionary theories of these and other pre-Darwinian Muslim scholars did influence some Western pre-Darwinian scholars and even Darwin himself (e.g., Guessoum, 2011). A careful comparison between the pre-Darwinian Muslim and Western literature is crucial to analyse in detail whether there was such a transmission of transmutational ideas from Muslim scholars to the West. However, as noted above, even if this was not the case, it does not mean that their ideas are any less relevant, because the innovative or accurate value of a scientific idea should not be measured by how useful it has been to Western thought or science, as if the latter were the pinnacle of evolution, the center of the world.

CHAPTER 4

BRINGING REALITY TO SCIENCE AND SOCIETY

"It does not matter how slowly you go as long as you do not stop."
(Confucius)

Works such as those presented in this book, have been changing the mindset of Western scholars and the broader public. But this transformation has been too slow. That is, until now. At the time of putting this book together, an interesting paper entitled: *People of the Book: Empire and Social Science in the Islamic Commonwealth Period*, was published by Musa Al-Gharbi. The paper shows that not only have the anatomical discoveries or transmutation ideas of Muslim scholars been ignored for many centuries, but so have the social sciences:

> Modernity came to be viewed as a universal process which, while birthed in Europe, would soon swallow all other societies and cultures as well. Every religion would undergo some version of reformation and enlightenment (or be lost to history). Every society would become secular, industrial, urban and capitalist (e.g. Weber, [1905] 2001). Although this process would generally be as brutal, destructive and disorienting for others as it was for Europe, these birthing pangs were predicted to eventually give way to an age of unprecedented worldwide peace and prosperity. Looking to other historical and cultural contexts increasingly became a means through which people in the West demonstrated their superiority over others and justified their eschatology of progress (Connell, 1997). 'The West' was thereby placed, in a sense, beyond good and evil. There could be no blame in destroying communities and cultures which were already destined for extinction. Although perhaps tragic to observe, it was ultimately a *kindness* to efficiently eliminate socio-cultural

phenotypes which were unfit for survival in the 'modern' world, or to accelerate mankind's evolution by reconstructing the exotic and primitive 'other' in the image of the 'modern' West. Operating under such auspices, [19th century Western] social science would go on to play a critical role in justifying imperialism (Ake, 2000), colonialism (Rhode, 2015), eugenics (Leonard, 2016), and genocide (Bauman, 2006).

Many have argued that the colonial and imperial origins of Western social science continue to structure its theory and practice today. Territoriality around scholarly disciplines is held to reflect an imperial mindset (Steinmetz 2007). Researchers continue to value forms of knowledge useful for administration, surveillance or dominance of subaltern populations—with 'social problems' defined largely in terms of these populations (Magubane, 2013). 'Insider' and 'outsider' groups continue to be studied and discussed in asymmetrical ways, with the sociological 'lens' turned primarily towards the latter (Latour, 1993). Social research continues to be largely extractive: data is taken from populations to be utilized and analyzed by outsiders. The preferences, priorities and perspectives of those from whom information is collected are often not taken seriously, or in any case, tend to be subordinated to the needs and desires of researchers (Connell, 2018). Aspirations towards empiricism, objectivity, rationality, universality and positivism are held to privilege values and worldviews prevalent among those who dominate the social order. Meanwhile, work by and for people from historically marginalized and disadvantaged groups tends to be viewed as less rigorous or significant (Go, 2020: Kemple and Mawani, 2009). These and other unsavoury features are held to be consequences of the imperial origins of Western social science.

In a nutshell, Al-Gharbi summarises the circular reasoning, self-reinforcing narratives, and links between ethnocentrism, politics, and societal and scientific biases, not only within social sciences but actually within the sciences as a whole. Similarly, this has applied not to a single group of peoples, but to many types of groups; the neglect, humiliation, and antagonism towards their habits, cultures, and even their scientific discoveries or broader ideas about the natural world. The untold stories about Muslim scholars is just one example amongst many, but it is a powerful one because the Muslim world has been a key target of Western scholars and the broader public, to showcase how 'superior' the Christian West is to 'others'. Minimising what Muslim scholars have done, either because they were 'believers', or because they did not truly use a 'scientific method' such as 'ours', or because what they said was not 'exactly similar' to what Western scholars, such as Darwin said, is part of those narratives about how 'superior' Westerners are. Strikingly, some of the points raised by Al-Gharbi concerning this issue are very similar to the ones we made earlier in this book, which were written before Al-Gharbi's paper was published. For instance, we noted that science was clearly not invented by Europeans a few centuries ago, a point also made by Al-Gharbi:

> Ibn Al-Haytham [a Muslim mathematician born in 965 in Basra, Iraq, who died in Cairo in 1040] was among the first to articulate a systematic approach for empirical research, now referred to as a 'scientific method': 1. Identify a problem; 2. Formulate a hypothesis about the answer to that problem; 3. Test the hypothesis through experimentation... if valid, he argued, these experiments and results should be reproducible; 4. Analyse the result rigorously by means of logic and mathematics; 5. Come to a conclusion about what (if anything) can be taken away from the experiment; 6. Publish the findings if they seem worthy—along with the methods and data—to be critiqued, refined and built upon by others.

In fact, even before Ibn Al-Haytham and his 'scientific method', scientific enquiries about the natural world had been undertaken by scholars in a number of societies, including Aristotle, anatomists in Ancient China, as well as the Mayans and Egyptians who studied and predicted the movements of the stars, and Asian mariners who could navigate more than 4,000 miles of sea to reach Madagascar, and so on. Another point made by Al-Gharbi that is also strikingly similar to what we have discussed, is about Muslim scholars being mainly seen as passive players and, thus, mostly credited for 'preserving' texts from Ancient Greece and the Roman Empire, which were then used millennia later by other Europeans. It is indeed quite revealing that such narratives were applied not only to natural sciences, as we have seen in this book, but also to the social sciences and various other fields of science and knowledge, as emphasised by Al-Gharbi:

Curiously, although empire is a social formation that repeats across historical, geographical and cultural contexts, the relationship between empire and social science tends to be discussed exclusively in terms of the 'modern' world and the West, as though this connection was historically *sui generis*. Given the much-discussed utility of social science *for* imperial regimes, the apparent lack of other historical cases seems to suggest that the creation of Western social science is equivalent with the creation of social science per se: perhaps the connections between empire and social science are not explored in other cases because no prior empire *had* social science at its disposal. In fact, centuries prior to the advent of Western social science, Islamic empires had produced their own 'science of society'. Exploring the character of this alternative imperial social research enterprise may help us gain leverage on the extent to which criticised attributes of Western social science are actually products of imperialism per se or are, perhaps, the result of other factors.

Yet although there is a corpus of excellent scholarship on Muslim contributions to mathematics or the material sciences, there are virtually no English-language resources detailing the breadth, depth and character of the Islamic social research program and how it fit into the broader imperial scientific enterprise. This gap in the literature can be attributed to many factors: The selection of works to be translated into Western languages or explored by Western scholarship has typically been governed by narrow instrumentalism and Whiggish historiography. For instance, Islamic scholars have been credited for 'preserving' Greek texts, allowing them to ultimately be 'restored' to the West; they have been praised for their 'early' contributions to mathematics, medicine and the physical sciences. These priorities: 'recovering' Greek philosophy and importing advances in science and technology, defined initial translation efforts by scholars like Michael Scot and Adelard of Bath towards the tail end of the Crusades (Lyons, 2010). In the lead up to the Enlightenment, there was another wave of translation work focused on the Qur'an, tafsir (Qur'anic exegesis) and Islamic theology and jurisprudence, undertaken by Christian scholars who were seeking to understand Islam in order to refute it, to demonstrate the superiority of Christianity and Western civilization and, ultimately, to convert the Arabs (Bevilaqua, 2018); the 'orientalist' scholarly enterprise of the colonial and imperial period served a similar set of ends (Said, 1978).

More generally, contributions to mathematics, medicine and material sciences were prized by Western scholars for their perceived practical significance. Social science, however, is and has been consistently held in lower esteem. Indeed, some of the figures discussed here for their pioneering work in the social sciences are much better known for their innovations in the realms of medicine, mathematics, physics, astronomy, etc. despite often having written just as much (or more) on social issues. The existing literature on Islamic scholars' contributions to social science is usually centered on specific disciplines or subfields, and oriented towards highlighting how these scholars either anticipated or contributed to the emergence of Western social science. That is, the 'modern' West is still the locus around which this scholarship is oriented; the work of these scholars is viewed as significant in virtue of some established relation with the 'mature' social sciences we are familiar with today.

It is true that *'the modern West is still the locus around which this scholarship is oriented'* but also that this orientation, even within the 'modern' West itself, is still mainly oriented to reinforce longstanding and inaccurate Western-centric, racist and sexist narratives. For instance, while we were putting together this book there were two independent studies conducted which showed just how baseless and biased the depictions of our past were. The major problem is that those depictions are evident not only in scientific papers and textbooks, which a few scientists see and are made to feel comfortable with their 'superior' society, culture, colour, or gender. Such depictions can be seen everywhere, including scientific museums, visited by millions of kids every year, as well as educational material, documentaries and TV shows that are seen and internalised by hundreds of millions of kids worldwide.

One of those papers, entitled *Visual Depictions of Our Evolutionary Past*, was published by one of us (Diogo) and three colleagues, in February 2021. It empirically shows that the reconstructions of human fossils, as famous as Lucy and the Taung Child (**Fig. 4.1**), are not often based on a deep knowledge of the muscles of our living relatives, the chimpanzees and gorillas, nor use robust technical methods. Instead, although to a lesser extent than during the apogee of scientific racism in the 19th and early 20th centuries, such depictions continue to be highly influenced by factually inaccurate ethnocentric, racist and/or sexist narratives. This is because the artists doing these reconstructions mostly do it either 'by eye', or basing them on what they know about the muscles of our species. Thus, because our species has been around for 250 to 300 thousand years while Lucy is approximately 3 million years older, a substantial portion of what they produce is subjective and, therefore, prone to influence from their subconscious—or conscious, in some cases—biases.

Fig. 4.1. Casts of facial reconstructions of Lucy (left) and the Taung child (right). To highlight how the use of different models is crucial for the final image, the authors reconstructed Lucy's tone to appear more similar to that of bonobos, whereas the Taung child's tone is more similar to that of anatomically modern humans native to South Africa.

Even more problematic is the fact that when the general public sees such depictions in museums or documentaries, they take them as *reality*. The same thing happens when they read in textbooks or see in documentaries and TV shows that science only appeared centuries ago in Europe, or that Muslim scholars never discussed any type of transmutation ideas or made any anatomical discoveries. A particularly powerful example provided in a 2021 press release concerns one of the most renowned natural history museums: the Smithsonian National Museum of Natural History in Washington, D.C. Currently, there is no scientific evidence supporting a progressive 'whitening of the skin' within humans before they left Africa because a lighter skin pigmentation was chiefly only selected in non-African regions with high or low latitudes. However, some displays in the Smithsonian—which is visited every year by millions of people, some of whom will take such depictions as representing the reality about human evolution—show such a progressive 'whitening of the skin'.

This progressive 'whitening of the skin' is seen in other Western museums, textbooks, media images and a plethora of websites (see **Fig. 4.2**), and is, of course, just one more 'modern' version of the longstanding chain of life (**Fig. 1.9**) notion that there is 'progress' in the natural world, with humans, and in particular European 'white' humans, being at the pinnacle of that progress. In fact, what is clearly implicit in both **Figs. 1.9** and **4.2**, and countless similar images that have been made, seen, copied, and disseminated since the time of Ancient Greece, is that one is only at the pinnacle of the chain of being or the ladder-of-life if one has 'white' skin *and* ideally if the colour of your hair is also not dark, as shown in **Fig. 4.2**. If you are not paleskinned, then you are far down that ladder, at a 'lower', 'inferior' position, and you should be well aware of that, and behave accordingly.

Fig. 4.2. Biased science and popular culture: marching towards evolutionary 'progress', leading to the 'pinnacle of evolution'—'white' Western males.

The second paper, published by Garcia-Campos and colleagues less than a month after the first one in March 2021, was entitled, *Indicators of Sexual Dimorphism in Homo antecessor Permanent Canines*. It revealed that one of the most famous fossils in Europe, found in the Sierra de Atapuerca in Spain, and belonging to *Homo antecessor*, is of a girl aged between 9 and 11 years (**Fig. 4.3**), and not from a boy, as had been initially believed. The new evidence came from an analysis of the canine teeth of the previously named 'boy from the Gran Dolina', in reference to the title of an emblematic book by José María Bermúdez de Castro. What is particularly telling in that paper is that its authors, who include Bermúdez de Castro, recognised that there was no scientific reason for assertion that the remains of the fossil were of a boy: 'it arose randomly... when José María [Bermúdez de Castro] decided to make the book, he chose this masculine name, but for no specific reason... it has been necessary to wait for these new techniques to be able to know the sex with certainty', explained García-Campos, one of the co-authors. The problem is that such assertions were made frequently, with almost all depictions of our past showing me— the very term 'cavemen' seems to imply that women basically were absent, or completely passive. The few cases in which women are depicted, they are shown as passive mothers, and not as active inventors or cave painters or food gatherers, despite the fact that available anthropological data shows that some women did each of these things.

Fig. 4.3. Analysis of the teeth of the "boy from the Gran Dolina" shows that the fossil was actually... a girl.

In the case of that paper, the study showed that this girl was likely involved in a process of interaction between groups, thus, causing us to rethink the role of women in pre-historical societies. As stated by the authors, it 'questions the traditional gender roles that are still preserved in which the woman is at home and the man at work' and helps 'to change the collective imagination of the female in the cave with two young or tanning skins' and to 'show us that women participated in hunting work and in disputes over territory'. Pressure is thus—finally—mounting on scholars and historians of science to be more active in dealing with the longstanding racist, sexist and Western-centric biases that influence many works, particularly within the two fields that were the 'father and mother' of scientific racism—biology and anthropology, thus including human comparative anatomy and human evolutionary biology. Unless those biases can be demonstrated and emphasised—as our book and these two recent papers have done—and subsequently corrected, the factual inaccuracies and biases of the past that continue to plague science, education and the media will condemn our kids to propagate and internalise them, and to impose or suffer the same types of discrimination, oppression and atrocities of the past.

REFERENCES AND SUGGESTED FURTHER READING

"The rapidity of the progress made by Islam in the sciences, arts, industry, and commerce, and all the refinements of civilized life, is almost as amazing as the rapidity of its conquest."
(Henri Pirenne)

Abdel-Halim RE, Abdel-Maguid TE. 2003. The functional anatomy of the uretero-vesical junction - a historical review. *Saudi Med J* 24:815-819.

Achtner W. 2009. The evolution of evolutionary theories of religion. In *The biological evolution of religious mind and behavior* (Voland E, Schiefennhovel W, eds). Springer, New York. p. 257-273.

Ahmedullah M. 2014. *Ibn Khaldun and Karl Marx: five centuries of history and two civilisations apart, yet remarkably similar*. http://alochonaa.com, 22 October 2014.

Alakbarli F. 2001. A 13th century Darwin? Tusi's Views on Evolution. *Azerbaijan Int* 9:48-49.

Alghamdi M, Ziermann JM, Diogo R. 2017. An untold story: the important contributions of Muslim scholars for the understanding of human anatomy. Anat Rec 300:986-1008.

Al-Gharbi M. 2021. People of the book: empire and social science in the Islamic Commonwealth period. *Socius* 7: 10.1177/23780231211021200

Al-Ghazal SK. 2007. Ibn Al-Nafis and the discovery of the pulmonary circulation. *Islamic Medicine On-line*: http://www. islamicmedicine.org.

Al-Qattan MM. 2005. History of anatomy of the hand and upper limb. *J Hand Surg* 31:502.

Al-Razi, Abu-Bakr Muhammad ibn Zakariya. ca. 1400-1500. *Kitab al-Mansouri fi al-Tibb (The book on medicine dedicated to al-Mansur)*. Library of congress: https://www.wdl.org/en/item/4276/#q=al-razi&qla=en

Al-Razi, Abu-Bakr Muhammad ibn Zakariya. ca. 1674. *Al-Hawi fi Al-Tibb (The Comprehensive Book in Medicine)*. Library of congress: https://www.wdl.org/en/item/9715/

Amr S, Tbakhi A. 2007. Abu Bakr Muhammad ibnZakariya Al Razi (Rhazes): Philosopher, Physician and Alchemist. *Ann Saudi Med* 27:305-307.

Andreassen R. 2014. Danish perceptions of race and anthropological science at the turn of the twentieth century. In *The invention of race - scientific and popular representations* (Bancel N, David T, Thomas D, eds.). Taylor & Francis, London, p. 117-129

Ardalan MR, Shoja MM, Tubbs RS, Eknoyan G. 2007. Diseases of the kidney in medieval Persia - the Hidayat of Al-Akawayni. *Nephrol Dial Transpl* 22:3413-3421.

Asghar A, Wiles FR, Alters B. 2010. The origin and evolution of life in Pakistani high school. Biol J Biol Educ 44:65-71.

Asghar A, Wiles JR, Alters B. 2007. Discovering international perspectives on biological evolution across religions and cultures *Int J Diversity* 6:81-89.

Asghar A. 2013. Canadian and Pakistani Muslim teachers' perceptions of evolutionary science and evolution education Evolution: Education and Outreach 6:1-12.

Asma ST. 2009. *On monsters - an unnatural history of our worst fears*. Oxford University Press, Oxford.

Attar S. 2007. *The vital roots of European enlightenment: Ibn Tufayl's influence on modern Western thought*. Lexington Books, Washington DC.

Ayala F. *Darwin's gift to science and religion*. Joseph Henry Press, Washington DC.

Bakhtiar L. 1999. *The Canon of Medicine (Al-Qanun fi'l-tibb)*. Kazi Publications Inc, Chicago.

Bancel N, David T, Thomas D, eds. 2014. *The invention of race - scientific and popular representations*. Taylor & Francis, London.

Barsanti G. 2009. *L'uomo dei boschi. piccola storia delle grandi scimmie da Aristotele a Darwin*. Università La Sapienza, Roma.

Bayrakdar M. 1983. Al-Jahiz and the rise of biological evolution. *Islamic Quarterly* 27:307-315.

Bering J. 2011. *The belief instinct - the psychology of souls, destiny, and the meaning of life*. W. W. Norton & Company, New York.

Besteiro JMF, Morales CAD, eds. 1987. *Colliget*, in *Aristotelis opera cum Averrois commentariis*. Venise apud Junctas, 1562–1574 (réimpr. Frankfurt: Minerva; for the Arabic, see Ibn Rušd, Kitāb al-Kullīyyāt fī l-ṭibb). Consejo Superior de Investigaciones Científicas-Escuela de Estudios árabes de Granada, Madrid.

Bethencourt F. 2013. *Racisms - from the crusades to the twentieth century.* Princeton University Press, Princeton.

Black E. 2003. *War against the weak - eugenics and America's campaign to create a master race.* Four Walls Eight Wondows, New York.

Blumberg MS. 2009. *Freaks of nature: what anomalies tell us about development and evolution.* Oxford University Press, New York.

Blume M. 2009. The reproductive benefits of religious affiliation. In *The biological evolution of religious mind and behavior* (Voland E, Schiefenn-hovel W, eds). Springer, New York. p. 117-149.

Blumenbach F. 1804. *De l'unite du genre humain.* Allut, Paris.

Boetsch G, Blanchard P. 2014. From cabinets of curiosity to the "Hottentot Venus": a long history of human zoos. In The invention of race - scientific and popular representations (Bancel N, David T, Thomas D, eds). Taylor & Francis, London, p. 185-194.

Bonadeo CM. 2013. *Abd Al-Laṭīf Al-Baġdādī's Philosophical Journey: From Aristotle's Metaphysics to the 'Metaphysical Science'.* Brill, New York.

Bondeson J. 1997. *A cabinet of medical curiosities.* W. W. Norton & Company, New York.

Bondeson J. 2000. *The two-headed boy, and other medical curiosities.* Cornell University Press, Ithaca.

Bonner JT. 2013. *Randomness in evolution.* Princeton University Press, Princeton.

BouJaoude S, Asghar A, Wiles JR, Jaber L, Sarieddine D, Alters B. 2011. Biology professors' and teachers' positions regarding biological evolution and evolution education in a Middle Eastern society. *Int J Sci Educ* 33:979-1000.

Boulter, M. 2013. *Scienceandartblog: Early Islam.* http://scienceandartblog.com/2013/09/12/early-islam/.

Bowler PJ. 1987. *Theories of human evolution - a century of debate, 1844-1944.* John Hopkins University Press, Oxford.

Bowler PJ. 2013. *Darwin deleted.* University of Chicago Press, Chicago.

Bowler PJ. 2017. Alternatives to Darwinism in the early twentieth century. In *The darwinian tradition in context: research programs in evolutionary biology* (Delisle RG, ed.). Springer, New York, p. 195–218.

Brattain M. 2007. Race, racism, and anti-racism: UNESCO and the politics of presenting science to the postwar public. *Amer Hist Rev* 112:1386-1413.

Brune M. 2009. On shared psychological mechanisms of religiousness and delusional beliefs. In The biological evolution of religious mind and behavior (Voland E, Schiefennhovel W, eds). Springer, New York, p. 217-228.

Buklijas T, Gluckman PD. 2013. From Evolution and Medicine to Evolutionary Medicine. In *The Cambridge Encyclopedia of Darwin and Evolutionary Thought* (Ruse M, ed). Cambridge University Press, Cambridge, p. 505-514.

Butler FP. 2012. *Evolution without darwinism - the legacy of Stephen Jay Gould*. CreateSpace, New York.

Campbell D. 2013. *Arabian Medicine and its Influence on the Middle Ages: Vol. 2. Arabian medicine and its influence on the Middle Ages*. Routledge, New York.

Campbell R, Vinas G, Henneberg M, Diogo R. 2021. Visual depictions of our evolutionary past: a broad case study concerning the need for quantitative methods of soft tissue reconstruction and art-science collaborations. *Frontiers Ecol Evol* 26:10.3389/fevo.2021.639048.

Canizares-Esguerra J. 2006. *Nature, empire and nation - explorations of the history of science in the Iberian world*. Stanford University Press, Stanford.

Carroll SB. 2020. *A series of fortunate events: chance and the making of the planet, life, and you*. Princeton University Press, Princeton.

Casserius I. 1600-1601. *De Vocis Auditus Que Organis Historia Anatomica*. Ferrariae, Venice.

Castro JM. 2021. Indicators of sexual dimorphism in *Homo antecessor* permanent canines. *J Anthropol Sci*: 10.4436/JASS.99001.

Catholic Church. 1983. *Code of canon law, Latin-English edition*. Catholic Church, Vatican.

Cerdá-Olmedo E. 2008. Ibn Tufayl (Abentofail) and the origins of scientific method. *Eur Rev* 16:159-167.

Church of England. 1844. *The Book of common prayer: printed by Whitchurch, March 1549; commonly called The first book of Edward VI.* William Pickering, London.

Cloquet J. 1821-1831. *Anatomie de 'homme.* Charles-Philibert, Paris.

Cole FJ. 1975. *A history of comparative anatomy - from Aristotle to the eighteenth century.* Dowe Publications, New York.

Contadini A. 2007. *Arab painting: text and image in illustrated Arabic manuscripts.* Brill, New York.

Coontz S. 2005. *Marriage, a history: how love conquered marriage.* Penguin Books, New York.

Corbey RHA, Theunissen B, eds. 1995. *Ape, man, apeman: changing views since 1600.* Leiden University, Leiden.

Corbey RHA. 2005. *The metaphysics of apes: negotiating the animal-human boundary.* Cambridge University Press, Cambridge.

Cowdry EV. 1921. A comparison of ancient Chinese anatomical charts with the 'fünfbilderserie' of sudhoff. Anat Rec 22:1-25.

Coyne J. 2016. Why do some scientists always claim that evolutionary biology needs urgent and serious reform? Blogpost: https://whyevolutionistrue.wordpress.com/2016/12/26/why-are-scientists-always-saying-that-evolutionary-biology-needs-urgent-and-serious-reform/.

Crews F. 2017. *Freud: the making of an illusion.* Metropolitan Books, New York.

Croutier AL. 1991. Harem: the world behind the veil. Abbeville Press, New Work.

Csoka AC. 2016. Innovation in medicine: Ignaz the reviled and Egas the regaled. *Med Health Care Philos* 19:163-168.

Cunningham A. 1997. *The anatomical Renaissance: the resurrection of the anatomical projects of the ancients.* Aldershot Scolar Press, London.

Cuvier G. 1797. *Tableau elementaire de 'histoire naturelle des animaux.* Bandouin, Paris.

DailyNews. 2010. Muslims - Founders of great libraries in history. http://archives.dailynews.lk/2010/10/15/fea26.asp.

Dajani R. 2015. Why I teach evolution to Muslim students. *Nature* 520:409.

Dalfardi B, Heydari M, Golzari SE, Nezhad M, Sadat G, Hashempur MH. 2014b. Al-Baghdadi's description of venous blood circulation. *Int J Cardiol* 174:209–210.

Dalfardi B, Mahmoudi Nezhad GS, Mehdizadeh A. 2014a. How did Haly Abbas look at the cardiovascular system?. *Int J Cardiol* 172:36-39.

Dalfardi B, Yarmohammadi H. 2014. Al-akhawayni and early differentiation between nerves and tendons. *J Hand Surg* 39:808.

Daneshfard B, Dalfardi B, Nezhad GSM. 2014b. Ibn Al-Haytham (965–1039 AD), the original portrayal of the modern theory of vision. *J Med Biogr*: 10.1177/0967772014529050.

Daneshfard B, Yarmohammadi H, Dalfardi B. 2014a. The origins of the theory of capillary circulation. *Int J Cardiol* 172:491–492.

Darwin C. 1859. On the origins of species by means of natural selection. Murray, London.

Darwin C. 1871. *The descent of man, and selection in relation to sex.* J. Murray, London.

Daston L, Park K. 1998. *Wonders and the order of nature, 1150-1750.* Zone Books, New York.

De Queiroz A. 2014. *The monkey's voyage - how improbable journeys shaped the history of life.* Basic Books, New York.

De Wall F. 2016. *Are we smart enough to know how smart animal are?* WWW Norton & Company, New York.

De Wall F. 2019. *Mama's last hug: animal emotions and what they tell us about ourselves.* WWW Norton & Company Inc, New York.

Delisle RG, ed. 2017a. The darwinian tradition in context - research programs in evolutionary biology. Springer, New York.

Delisle RG. 2007. Debating humankind's place in nature, 1860–2000: the nature of paleoanthropology. Pearson Prentice Hall, Upper Saddle River NJ.

Delisle RG. 2017b. From Charles Darwin to the evolutionary synthesis: weak and diffused connections only. In *The Darwinian tradition in context: research programs in evolutionary biology* (Delisle RG, ed.). Springer, New York, p. 133–167.

Delisle RG. 2019. *Charles Darwin's incomplete revolution - the origin of species and the static worldview.* Springer, New York.

Depew DJ. 2017. Darwinism in the twentieth century: productive encounters with saltation, acquired characteristics, and development. In *The Darwinian tradition in context: research programs in evolutionary biology* (Delisle RG, ed.). Springer, New York, p. 61–68.

DeSesso JM. 2019. The arrogance of teratology: a brief chronology of attitudes throughout history. *Birth Defects Res* 111:123-141.

Diamandopoulos A, Goudas C. 2007. Human and ape: the legend, the history and the DNA. *Hippokratia* 11:92.

Diamond J. 1999. *Guns, germs, and steel: the fates of human societies.* W. W. Norton & Company, New York.

Diamond J. 2012. *The world until yesterday: what we can learn from traditional societies?* Penguin Books, New York.

Diamond J. 1991. *The rise and fall of the third chimpanzee.* Hutchinson Radius, London.

Diamond, J. 2005. *Collapse: how societies choose to fail or succeed.* Viking Press, New York.

Diogo R, Abdala V. 2010. *Muscles of vertebrates - comparative anatomy, evolution, homologies and development.* CRC Press, New Hampshire.

Diogo R, Bello-Hellegouarch G, Kohlsdorf T, et al. 2016a. Comparative myology and evolution of marsupials and other vertebrates, with notes on complexity, Bauplan, and «Scala Naturae». *Anat Rec* 299:1224-1255.

Diogo R, Guinard G, Diaz R. 2017b. Dinosaurs, chameleons, humans and Evo-Devo-Path: linking Étienne Geoffroy's teratology, Waddington's homeorhesis, Alberch's logic of 'monsters', and Goldschmidt hopeful 'monsters'. *J Exp Zool B* 328:207-229.

Diogo R, Noden D, Smith CM, et al. 2016b. *Learning and understanding human anatomy and pathology: an evolutionary and developmental guide for medical students.* Taylor & Francis, Oxford.

Diogo R, Pastor JF, Hartstone-Rose A, et al. 2014. *Baby Gorilla: photographic and descriptive musculoskeletal atlas of the skeleton, muscles and internal organs - including CT scans and comparisons to other gorillas and primates.* Taylor & Francis, Oxford.

Diogo R, Potau JM, Pastor JF, et al. 2010. *Photographic and descriptive musculoskeletal atlas of Gorilla - with notes on the attachments, variations, innervation, synonymy and weight of the muscles.* Taylor & Francis, Oxford.

Diogo R, Potau JM, Pastor JF, et al. 2012. *Photographic and descriptive musculoskeletal atlas of gibbons and siamangs (Hylobates) - with notes on the attachments, variations, innervation, synonymy and weight of the muscles.* Taylor & Francis, Oxford.

Diogo R, Potau JM, Pastor JF, et al. 2013a. *Photographic and descriptive musculoskeletal atlas of chimpanzees (Pan) - with notes on the attachments, variations, innervation, synonymy and weight of the muscles.* Taylor & Francis, Oxford.

Diogo R, Potau JM, Pastor JF, et al. 2013b. *Photographic and descriptive musculoskeletal atlas of orangutans (Pongo) - with notes on the attachments, variations, innervation, synonymy and weight of the muscles.* Taylor & Francis, Oxford.

Diogo R, Shearer B, Potau JM, et al. 2017c. *Photographic and descriptive musculoskeletal atlas of bonobos, with notes on the attachments, variations, innervation, synonymy and weight of the muscles.* Springer, New York.

Diogo R, Wood B. 2011. Soft-tissue anatomy of the primates: phylogenetic analyses based on the muscles of the head, neck, pectoral region and upper limb, with notes on the evolution of these muscles. J Anat 219:273–359. Diogo R, Wood B. 2012. Comparative anatomy and phylogeny of primate muscles, and human evolution. CRC Press, New Hampshire.

Diogo R, Wood B. 2013. The broader evolutionary lessons to be learned from a comparative and phylogenetic analysis of primate muscle morphology. Biol Rev 88:988-1001.

Diogo R, Ziermann JM, Linde-Medina M. 2015. Is evolutionary biology becoming too politically correct? A reflection on the scala naturae, phylogenetically basal clades, anatomically plesiomorphic taxa, and "lower" animals. *Biol Rev* 90:502-521.

Diogo R. 2010. Comparative anatomy, anthropology and archaeology as case studies on the influence of human biases in natural sciences: the origin of 'humans', of 'behaviorally modern humans' and of 'fully civilized humans'. *Open Anat J* 2:86-97.

Diogo R. 2017a. *Evolution driven by organismal behavior - a unifying view of life, function, form, trends and mismatches.* Springer, New York.

Diogo R. 2018a. Links between the discovery of primates and anatomical comparisons with humans, the chain of being, our place in nature, and racism. *J Morphol* 279:472-493.

Diogo R. 2018b. Where is, in 2017, the Evo in Evo-Devo (Evolutionary Developmental Biology)? *J Exp Zool B* 330:15-22.

Diogo R. 2019. Sex at Dusk, Sex at Dawn, selfish genes: how old-dated evolutionary ideas are used to defend fallacious misogynistic views on sex evolution. *J Soc Sci Humanit* 5:350-367.

Diogo R. In preparation. *Racism, sexism, and Darwin's idolization - and their tragic scientific and societal repercussions until today.*

Diogo R., Molnar JL, Wood B. 2017a. Bonobo anatomy reveals stasis and mosaicism in chimpanzee evolution, and supports bonobos as the most appropriate extant model for the common ancestor of chimpanzees and humans. *Sci Rep* 7:608.

Diogo, R. 2017b. Etho-eco-morphological mismatches, an overlooked phenomenon in ecology, evolution and Evo-Devo that supports ONCE (Organic Nonoptimal Constrained Evolution) and the key evolutionary role of organismal behavior. *Front Ecol Evol - EvoDevo*:10.3389.

Draper JW. 1875. History of the conflict between religion and science. D. Appleton, New York.

Draper JW. 1876. History of the intellectual development of Europe. D. Appleton, New York.

Duke D. 1998. *My awakening: a path to racial understanding*. Free Speech Press, Covington.

Dyble M, Thorley J, Page AE, *et al.* 2019. Engagement in agricultural work is associated with reduced leisure time among Agta hunter-gatherers. *Nature Hum Behav*: s41562-019-0614-6.

Eldredge N, Gould SJ. 1972. Punctuated equilibrium: an alternative to phyletic gradualism. In *Models in paleobiology* (Schopf TJM, ed.). Freeman, Cooper and Co., San Francisco, p. 82-115.

Eldredge N. 2014. *Extinction and evolution: what fossils reveal about the history of life*. Firefly Books, Toronto

Engelmeier H. 2016. *Der Mensch. Der Affe*. Böhlau Verlag, Köln.

Epstein GM. 2010. *Good without God - what a billion nonreligious people do believe*. Harper, New York.

Erb CM. 1998. *Tracking King Kong - a Holliwood icon of world culture*. Wayne State University Press, Detroit.

Fábrega H. 1997. *Evolution of sickness and healing*. University of California Press, Berkeley.

Fabrici G. 1600. *De Formato Foetu*. Embryo Project Encyclopedia (2008-08-27), ISSN, 1940-5030.

Fine C. 2017. *Testosterone Rex - myths of sex, science and society*. W. W. Norton & Company, New York.

Frey U. 2009. Cognitive foundations of religiosity. In *The biological evolution of religious mind and behavior* (Voland E, Schiefennhovel W, eds). Springer, New York, p. 229-241.

Fuentes A. 2017. Human niche, human behaviour, human nature. *Interface Focus* 7:20160136

Futuyma DJ. 2017. Evolutionary biology today and the call for an extended synthesis. *Interface Focus* 7: 20160145.

García-Campos C, Martinén-Torres M, Modesto-Mata M, Martín-Francés L, Martínez de Pinillos M, Bermúdez de

Garstang M. 2015. *Elephant sense and sensibility.* Academic Press, New York.

Golzari S, Khan Z, Ghabili K, Hosseinzadeh H, Soleimanpour H, Azarfarin R, Mahmoodpoor A, Aslanabadi S, Ansarin K. 2013. Contributions of medieval Islamic physicians to the history of tracheostomy. *Intern Anesth Res Soc* 116:1123-1132.

Goodall J. 1988. *In the shadow of man.* Houghton Mifflin Company, Boston.

Gottschall J. 2012. *The storytelling animal - how stories make us human.* Houghton Mifflin Hartcourt, New York.

Gould SJ. 1996. *Full House: the spread of excellence from Plato to Darwin.* Belknap Press, Cambridge.

Gould SJ. 2002. *The structure of evolutionary theory.* Belknap, Harvard.

Gould SJ. 2002. *The structure of evolutionary theory.* Harvard University Press, Cambridge.

Gould SJ. 1981. *The mismeasure of man.* W. W. Norton & Company, New York.

Gray J. 2013. *The silence of animals: on progress and other modern myths.* Farrar, Straus & Giroux, New York.

Gray J. 2018. *Seven types of atheism.* Penguin Books, London.

Green T. 2019. *A fistful of shells - West Africa from the rise of the slave trade to the age of revolution.* The University of Chicago Press, Chicago.

Greenblatt S. 2011. *The swerve - how the world became modern.* W. W. Norton & Company, New York.

Greenblatt S. 2017. *The rise and fall of Adam and Eve.* W. W. Norton & Company, New York.

Groves C. 2008. *Extended family: long lost cousins. A personal look at the history of primatology.* Conservation International, Arlington.

Guedron M. 2014. Panel and sequence: classification and associations in scientific illustrations of the human races (1770-1830). In *The invention of race - scientific and popular representations* (Bancel N, David T, Thomas D, eds). Taylor & Francis, London, p. 60-67.

Guerrini A. 2003. *Experimenting with humans and animals - from Galen to animal rights.* The Johns Hopkins University Press, Baltimore.

Guerrini A. 2015. *The Courtiers' Anatomists - Animals and Humans in Loius XIV's Paris.* The University of Chicago Press, Chicago.

Guessoum N. 2011. *Islam's quantum question: reconciling Muslim tradition and modern science.* IB Tauris, New York.

Gupta M, Prasad NG, Dey S, et al. 2017. Niche construction in evolutionary theory: the construction of an academic niche? *J Genet* 96:491-504

Haas R, Watson J, Buonasera T, et al. 2020. *Female hunters of the early Americas. Science Adv* 6: eabd0310.

Hameed S. 2008. Bracing for Islamic creationism. *Science* 322:1637-1638.

Hameed S. 2012. Walking the tightrope of the science and religion boundary. *Zygon* 47:337-342

Hannon E, Lewens T, eds. 2018. *Why we disagree about human nature.* Oxford University Press, Oxford.

Harari YN. 2017. *Homo Deus - a brief history of tomorrow.* Harper, New York.

Hare B, Woods V. 2020. *The survival of the friendliest - understanding our origins and rediscovering our common humanity.* Random House, New York.

Haught JE. 2000. *Science and religion in search for of cosmic purpose.* Georgetown University Press, Washington DC.

Hawi SS. 1974. Islamic naturalism and mysticism: a ahilosophic study of Ibn 'oTufayl's 'oHayy Bin Yaqòżaan. Brill New York.

Hazlewood N. 2001. *Savage: the life and times of Jemmy Button.* Thomas Dunne Books, New York.

Hehmeyer I, Khan A. 2007. Islam's forgotten contributions to medical science. *Can Med Assoc J* 176:1467-1468.

Henning BG, Scarfe AC, eds. 2013. *Beyond mechanism: putting life back into biology.* Lexington Books, Lexington.

Hewlett BS, ed. 2014. *Hunter-Gatherers of the Congo basin: cultures, histories, and biology of African pygmies.* Transaction Publishers, London.

Hewlett BS. 2014. Hunter-gatherer childhoods in the Congo basin. In *Hunter-Gatherers of the Congo Basin: Cultures, Histories, and Biology of African Pygmies* (Hewlett BS, ed.). Transaction Publishers, London, p. 245-275.

Hochschild AR. 2016. *Stranger in their own land - anger and morning on the American right.* The New Press, New York.

Hoff EV. 2004. A friend living inside me - the forms and functions of imaginary companions. *Imagination, Cognit & Personal* 24:151–189.

Hoffman D. 2019. *The case against reality: why evolution hid the truth from our eyes.* W.W. Norton & Company Inc., New York.

Hoffmeyer J. 2013. Why do we need a semiotic understanting of life. In *Beyond mechanism: putting life back into biology* (Henning BG, Scarfe AC, eds). Lexington Books, Lexington, p. 147-168.

Holland J. 2012. *A brief history of mysogyny - the world's oldest prejudice.* Constable & Robinson Ltd., London.

Hood B. 2013. *The self illusion - how the social brain creates identity.* Oxford University Press, Oxford.

Hood RW, Hill PC, Spilka B. 2009. *The psychology of religion - an empirical approach.* The Guilford Press, London.

Hoquet T. 2014. Biologization of race and racialization of the human: Bernier, Buffon, Linnaeus. In *The invention of race - scientific and popular representations* (Bancel N, David T, Thomas D, eds). Taylor & Francis, London, p. 17-32.

Hrdy SB. 2009. Mother and others - the evolutionary origins of mutual understanding. Belknap Press, Cambridge.

Huber E. 1931. *Evolution of facial musculature and expression.* The Johns Hopkins University Press, Baltimore.

Humboldt A. 1914. Views of the cordilleras and monuments of the indigenous peoples of the Americas. In Views of the Cordilleras and monuments of the indigenous peoples of the Americas: a critical edition, 2012 (Kutzinski VM, Ette O, eds). University of Chicago Press, Chicago, p. 1-370.

Huneman P, Walsh DM, eds. 2017. *Challenging the modern synthesis - adaptation, development, and inheritance.* Oxford University Press, Oxford.

Huxley TH. 1863. *Evidence as to man's place in nature.* Williams and Norgate, London.

Ibn Abbas, Ali Al-Majusi, ca. 1437. Ketab Kamel Al-Sena-a Al-Tebiah (The Complete Art of Medicine), known as Al-Ketab Al-Malki (The Royal Book). Library of congress: https://www.wdl.org/en/item/9716/

Ibn Al-Nafis, Abu Al-Hasan, Alā' Al-Dīn 'Alī ibn Abī Al-ḥazm Al-Qarashī Al-Dimashqi. *Mujaz Al-Qanun (The Commentary on Anatomy in Avicenna's Canon).* Library of congress: https://www.wdl.org/en/item/16193/

Ibn Ilyas, Mansur ibn Mohammad ibn Ahmad ibn Yousef. 1709. *Tasrih-i Mansuri (Mansur's Anatomy), known as Tashrih-i Badan-i Insan*

(Human Anatomy). Library of congress: https://www.wdl.org/en/item/9719/#q=منصوري&qla=en

Ibn Rushd, Abū Al-Walīd Muḥammad ibn Aḥmad. *Al-Kulliyat Fi Al-Tibb (Generalities or General Medicine)*. Library of congress: https://lccn.loc.gov/2002388049

Ibn Sina, Abu Ali Husain ibn Abdullah, ca. 1597. *Al-Qanun fi Al-Tibb (The Canon of Medicine)*. Library of congress: https://www.wdl.org/en/item/9718/#q=The+Canon+of+Medicine

Ibn Tufayl. 2009. *Ibn Tufayl's Hayy Ibn Yaqzan: A philosophical tale*. University of Chicago Press, Chicago.

Ibn-Khaldūn 1377. Al-Muqaddimah, Chapter 1 - sixth prefatory discussion. Muslim philosophy (Trans. F. Rosenthal, 1967).. Princeton University Press, Princeton.

Ihsanoglu E. 2004. *Science, technology and learning in the Ottoman empire; Western influence, local institutions and the transfer of knowledge*. Variorum Aldershot, Ashgate.

Iqbal M. 2007. *Science and Islam*. Greenwood, Westport.

Jablonka E, Lamb MJ. 2005. *Evolution in four dimensions - genetic, epigenetic, behavioral, and symbolic variation in the history of life*. MIT Press, Cambridge.

Janson HW, ed. 1952. Apes and Ape lore in the Middle Ages and the Renaissance. Warburg Institute University of London, London.

Jobs S, Mackenthun G. 2011. Embodiments of Cultural Encounters. Waxmann Verlag, Berlin.

Johnson MR. 2005. Aristotle on teleology. Clarendon Press, Oxford.

Johnson NA, Lahti DC, Blumstein DT. 2012. Combating the assumption of evolutionary progress: lessons from the decay and loss of traits. *Evol Educat & Outreach 5:128-138*.

Kasperbauer TJ. 2018. Subhuman - the moral psychology of human attitudes to animals. Oxford University Press, Oxford.

Kauffman SA. 2010. *Reinventing the sacred: a new view of science, reason, and religion*. Basic Books, New York.

Kaya V. 2012. Can the Quran support Darwin? An evolutionist approach by two Turkish scholars after the foundation of the Turkish Republic. *Muslim World* 102:357-370.

Kechichian JA. 2012. *The father of the theory of evolution*. Al Nisr Publishing LLC, Middle East and online. doi:http://gulfnews.com/about-gulfnews/al-nisr-portfolio/weekend-review/the-father-of-the-theory-of-evolution-1.1079209.

Kelly RL. 1995. *The foraging spectrum: diversity in hunter-gatherer lifeways*. Smithsonian Institution Press, Washington.

Kelly RL. 2013. *The lifeways of hunter-gatherers - the foraging spectrum*. Cambridge Press, Cambridge.

Kendy IX. 2016. *Stamped from the beginning - the definitive history of racist ideas in America*. Nation Books, New York.

Kevles DJ. 1995. *In the name of eugenics - genetics and the uses of human heredity*. Harvard University Press, Cambridge.

Khalili M, Shoja MM, Tubbs RS, Loukas M, Alakbarli F, Newman AJ. 2010. Illustration of the heart and blood vessels in medieval times. *Int J Cardiol* 143:4-7.

Khodadoust K, Ardalan M, Ghabili K, Golzari SE, Eknoyan G. 2013. Discourse on pulse in medieval Persia - the Hidayat of Al-Akhawayni (?–983AD). Int J Cardiology 166:289-293.

Kirkegaard EOW. 2019. Race differences: a very brief review. *Mankind Quart* 60.2:142-173.

Kirkham D. 2019. *Our shadowed world - reflections on civilization, conflict, and belief*. Cascade Books, Eugene.

Kteily N, Bruneau E, Waytz A, *et al*. 2015. The ascent of man: theoretical and empirical evidence for blatant dehumanization. *J Pers Social Psychol* 109:901-931.

Kuklick H, ed. 2008. *A New History of Anthropology*. Blackwell, Oxford.

Kull K. 2014. Adaptive evolution without natural selection. *Biol J Linn Soc* 112:287-294.

Kunz J. 2009. Is there a particular role for ideational aspects of religions in human behavioral ecology? In *The biological evolution of religious mind and behavior* (Voland E, Schiefennhovel W, eds). Springer, New York, p. 89-104.

Lagerkvist U. 2005. *The enigma of ferment - from the philosopher›s stone to the first biochemical Nobel prize*. World Scientific, Hackensack.

Lahti DC. 2009. The correlated history of social organization, morality, and religion. In *The biological evolution of religious mind and behavior* (Voland E, Schiefennhovel W, eds). Springer, New York, p. 67-88.

Laland K, Matthews B, Feldman MW. 2016. An introduction to niche construction theory. *Evol Ecol* 30:191-202.

Laland KN, Odling-Smee J, Turner S. 2014. The role of internal and external constructive processes in evolution. *J Physiol* 592:2413-2422.

Laland KN, Uller T, Feldman MW, et al. 2015. The extended evolutionary synthesis: its structure, assumptions and predictions. Proc R Soc Lon B 282:10.1098/rspb.2015.1019.

Landau, M. 1991. *Narratives on human evolution*. Yale University Press, New Haven.

Lee RB. 2018. Hunter-Gatherers and human evolution: new light on old debates. *Annu Rev Anthropol* 47:513-531.

Leeman RW. 2012. *The teleological discourse of Barack Obama*. Lexington Books, New York.

Lenoir T. 1982. *The strategy of life - teleology and mechanics in nineteenth-century german biology*. The University of Chicago Press, Chicago.

Leroi AM. 2003. *Mutants: on the form, varieties and errors of the human body*. Harper Collins, London.

Leroi AM. 2014. *The lagoon: how Aristotle invented science*. Bloomsbury Publishing, New York.

Leroi AM. 2014. *The Lagoon: How Aristotle invented science*. Bloomsbury, London.

Levin J. 2020. *Religion and medicine - a history of the encounter between humanity's two greatest institutions*. Oxford University Press, Oxford.

Levi-Strauss C. 2011. *Tristes tropiques*. Penguin Books, New New York.

Lewis HS. 2001. Boas, Darwin, science, and anthropology. *Curr Anthropol* 42:381-406.

Lieberman P. 1991. *Uniquely human: the evolution of speech, thought and selfless behavior*. Harvard University Press, Cambridge.

Lindberg DC. 2003. Medieval Islamic achievement in optics. *Opt Photonics News* 14:30-35.

Lindholm M. 2015. DNA dispose, but subjects decide -learning and the extended synthesis. *Biosemiotics* 8:4431-4461.

Linnaeus C. 1735. *Systema naturae*. Laurentius Salvius, Stockholm.

Loukas M, Hanna M, Alsaiegh N, Shoja MM, Tubbs RS. 2011. Clinical anatomy as practiced by ancient Egyptians. *Clin Anat* 24:409-415.

Lovejoy AO. 1936. *The great chain of being: a study of the history of an idea*. Harvard University Press, Cambridge.

Malik A, Ziermann JM, Diogo R. 2017. An untold story in biology: the historical continuity of evolutionary ideas of Muslim scholars from the 8th century to Darwin's time. *J Biol Educ*: 10.1080/00219266.2016.1268190.

Marshall JM. 2007. *The day the world ended at Little Bighorn - a Lakota history*. Penguin Books, London.

Martin D. 1984. *Primate Origins and Evolution*. Chapman and Hall, London.

Martínez-Sevilla F, Arqués M, Jordana X, et al. 2020. Who painted that? The authorship of Schematic rock art at the Los Machos rockshelter in southern Iberia. *Antiquity* 94:1133-1151.

Mayr E. 1976. *Evolution and the diversity of life: selected essays*. Harvard University Press, Cambridge.

Mayr E. 1982. *The growth of biological thought: Diversity, evolution, and inheritance*. Harvard University Press, Cambridge.

McBrearty S, Brooks AS. 2000. The revolution that wasn't: a new interpretation of the origin of modern human behavior. J Hum Evol 39:453-563.

McShea DW. 2012. Upper-directed systems: a new approach to teleology in biology. Biol & Philos 27:663–684.

Meijer MC. 2014. Cranial varieties in the human and orangutan species. In *The invention of race - scientific and popular representations* (Bancel N, David T, Thomas D, eds). Taylor & Francis, London, p. 33-47.

Meyerhof M. 1926. New light on Hunain ibn Ishaq and his period. *Isis* 8:685-724.

Montagu MFA. 1943. Edward Tyson, M.D., F.R.S., 1650-1708. *Mem Am Philos Soc* 20:1-488.

Morris D. 2013. *Monkey*. Reaktion Books Ltd, London.

Moser S. 1998. *Ancestral images - the iconography of human origins*. Cornell University Press, Ithaca.

Muazzam MG, Muazzam N. 1989. Important contributions of early Muslim period to medical science. II. Clinical sciences. *IMANA* 21:64-70.

Muslim Heritage - *Women dealing with health during the Ottoman reign*. http://www.muslimheritage.com/article/women-dealing-health-during-ottoman-reign/gallery/952.

Nabipour I. 2003. Clinical endocrinology in the Islamic civilization in Iran. *Int J Endocrinol Metab* 1:43-45.

Najjar J. 2010. From anesthetic sponge to nonsinking skull perforator, unitary work neurosurgery in the ancient Arabic and Islamic World. *World Neurosurg* 73:587-594.

Nasr SH. 1993. An introduction to Islamic cosmological doctrines. SUNY Press, New York.

Nee S. 2005. The great chain of being. *Nature* 435:429-429.

Newman AJ. 1998. Tashrih-e Mansuri: human anatomy between the Galenic and prophetic medical traditions. In *La Science Dans le Monde Iranien* (eds. Vesel Z, Beikbaghban H, Thierry B), pp. 253-271. Institut Francais de Recherche en Iran, Tehran.

Nidhami-i-Arudi-i-Samarqandi (12th Century). *The Chahar Maqala ("Four discourses") of Nidhami-i-Arudi-i-Samarqandi* Trans. Edward G. Browne, 1899). Stephen Austin and Sons, London.

Nisbet R. 1980. *History of the idea of progress*. Basic Books, New York.

Nixey C. 2017. *The Darkening age - the Christian destruction of the classic world*. Macmillan, London.

Odling-Smee FJ, Laland KN, Feldman MW. 2003. *Niche construction – the neglected process in evolution (Monographs in population biology 37)*. Princeton University Press, Princeton.

Omland KE, Cook LG, Crisp MD. 2008. Tree thinking for all biology: the problem with reading phylogenies as ladders of progress. *BioEssays* 30:854–867.

Owen R. 1830. On the anatomy of the orangutan (*Simia satyrus*, L.). Proc Zool Soc London 1830: 4-5.

Panese F. 2014. The creation of the 'negro' at the turn of the nineteenth century: Petrus Camper, Johan Friedrich Blumenbach, and Julien-Joseph Virey. In *The invention of race - scientific and popular representations* (Bancel N, David T, Thomas D, eds). Taylor & Francis, London, p. 48-59.

Park K. 2006. *Secrets of women: gender, generation and the origins of human dissection*. Zone Books, New York.

Perrault C. 1676. *Memoires pour servir a l'histoire naturelle des animaux*. De l'Imprimerie Royale, Paris.

Persaud TVN, Loukas M, Tubbs RS. 2014. *A history of human anatomy, 2nd ed*. Charles C Thomas, Springfield.

Persaud TVN, Loukas M, Tubbs RS. 2014. *A history of human anatomy, 2nd ed*. Charles C Thomas, Springfield.

Persaud TVN. 1984. *Early history of human anatomy: from antiquity to the beginning of the modern era*. Charles C Thomas, Springfield.

Persaud TVN. 1984. *Early history of human anatomy: from antiquity to the beginning of the Modern Era*. Charles C Thomas, Springfield.

Peterson A. 2001. *Being human - ethics, environment, and our place in the world*. University of California Press, Berkeley.

Piazza PV. 2019. *Homo Biologicus: Comment la biologie explique la nature humaine*. Albin Michel, Paris.

Pigliucci M, Müller GB, eds. 2010. *Evolution - the extended synthesis*. MIT Press, Cambridge.

Pigliucci M. 2017. Darwinism after the modern synthesis. In *The Darwinian tradition in context: research programs in evolutionary biology* (Delisle RG, ed.). Springer, New York, p. 94-104.

Pinker S. 2011. *The better angels of our nature: why violence has declined'*. Penguin Books, New York.

Pormann PE, Savage-Smith E. 2007. *Medieval Islamic Medicine*. Georgetown University Press, Washington DC.

Pratarelli ME, Chiarelli B. 2007. Extinction and overspecialization: the dark side of human innovation. *Mankind Quart* 48:83-98.

Pringle P. 2008. The murder of Nikolai Vavilov - the story of Stalin's persecution of one of the great scientists of the twentieth century. Simon & Schuster, New York.

Prioreschi P. 2006. Anatomy in Medieval Islam. *JISHIM* 5:2-6.

Pruetz JD, LaDuke TC. 2010. Reaction to fire by savanna chimpanzees (Pan troglodytes verus) at Fongoli, Senegal: conceptualization of "fire behavior" and the case for a chimpanzee model. Am J Phys Anthropol 141:646–650.

Prum RO. 2017. *The evolution of beauty: how Darwin's forgotten theory of mate choice shapes the animal world - and us*. Anchor Books, New York.

Puchner M. 2017. *The written world: the power of stories to shape people, history, civilization*. Penguin Random House LLC, New York.

Qala'aji A A. 1994. Al-Tibb min Al-Kitab wa-Al-Sunna (Medicine from the Holy Book and the Life of the Prophet) by Muwaffaq Al-Din Abd Al-Latif Al-Baghdadi. *Dar El-Marefah*, Lebanon: 246-258.

Radini A, Tromp M, Beach A, *et al.* 2019. Medieval women's early involvement in manuscript production suggested by lapis lazuli identification in dental calculus. *Science Adv* 5:eaau7126.

Ramsey G, Pence CH, eds. 2016. *Chance in evolution*. The University of Chicago Press, Chicago.

Reiss JO. 2009. *Not by design: retiring Darwin's watchmaker*. University of California Press, Berkeley.

Reno PL, Horton WE Jr, Lovejoy CO. 2013. Metapodial or phalanx? An evolutionary and developmental perspective on the homology of the first ray's proximal segment. *J Exp Zool B* 320:276-285.

Reynaud-Paligot C. 2014. Construction and circulation of the notion of 'race' in the nineteenth century. In *The invention of race - scientific and popular representations* (Bancel N, David T, Thomas D, eds). Taylor & Francis, London, p. 87-99.

Richards RJ. 2008. *The tragic sense of life: Ernst Haeckel* and the struggle over evolutionary thought. University of Chicago Press, Chicago.

Richert RA, Smith EI. 2009. Cognitive foundations in the development of a religious mind. In *The biological evolution of religious mind and behavior* (Voland E, Schiefennhovel W, eds). Springer, New York, p. 181-204.

Rigato E, Minelli A. 2013. The great chain of being is still here. *Evo Educ & Outreach* 6:18.

Riva A, Orrù B, Pirino A, *et al.* 2001. Iulius Casserius (1552-1616): the self-made anatomist of Padua›s golden age. *Anat Rec* 265:168–175.

Roger F. 2000. *Ancients and Moderns in the Medical Sciences. From Hippocrates to Harvey*. Aldershot-Burlington, Ashgate.

Rossano M. 2009. The African interregnum: the 'where', 'when', and 'why' of the evolution of religion. In *The biological evolution of religious mind and behavior* (Voland E, Schiefennhovel W, eds). Springer, New York, p. 127-141.

Rovelli C. 2017. *Reality is not what it seems - the journey to quantum gravity*. Riverhead books, New York.

Ruse M. 1993. Will the real Charles Darwin please stand up? *Quart Rev Biol* 68:225-231.

Ruse M. 1996. *Monad to man: the concept of progress in evolutionary biology*. Harvard University Press, Cambridge.

Ruse M. 2003. *Darwin and design - does evolution have a purpose?* Harvard University Press, Cambridge.

Ruse M. 2013. From organisms to mechanisms - and halfway back? In *Beyond mechanism: putting life back into biology* (Henning BG, Scarfe AC, eds). Lexington Books, Lexington, p. 409-430.

Ruse M. 2018. *On purpose*. Princeton University Press, Princeton.

Russel ES. 1916. *Form and function - a contribution to the history of animal morphology*. John Murray, London.

Russel GA. 2010. Chapter 6: after Galen Late Antiquity and the Islamic world. *J Headache Clin Neurol* 95:61-77.

Sabrah A. 1983. *Kita-b Al-Manazir (Book of Optics) by Abu Ali Al-Ḥasan ibn Al-Ḥasan ibn Al-Haytham*. Al Majlis al-Waṭanī lil-Thaqāfah wa-al-Funūn wa-al-Ādāb, Kuwayt.

Sagan C. 1997. *The demon-haunted world - science as a candle in the dark*. Ballantine Books, New York.

Saini A. 2017. *Inferior - how science got women wrong, and the new research that's rewriting the story*. Beacon Press, Boston.

Sakai T. 2007. Historical evolution of anatomical terminology from ancient to modern. *Anat Sci Int* 82:65-81.

Sánchez-Villagra M, van Schaik CP. 2019. Evaluating the self-domestication hypothesis of human evolution. *Evol Anthropol* 28:133-143.

Saniotis A. 2012. Islamic medicine and evolutionary medicine: a comparative analysis. J IMA 44.

Sapolsky RM. 2017. *Behave - the biology of humans at our best and worst*. Penguin Press, New York.

Savage-Smith E. 1995. Attitudes toward dissection in medieval Islam. *J Hist Med Allied Sci* 50:67-110.

Savage-Smith E. 1996. Medicine. In *Encyclopedia of the History of Arabic Science* (Rashed R., Ed), p. 948-953). Routledge, New York.

Savage-Smith E. 2005. *Mansur ibn Ilyas. Tashrih-i badan-i insan [Anatomy of the Human Body]*. http://www.nlm.nih.gov/exhibition/historicalanatomies/mansur_bio.html.

Săvoiu G. 2014. The impact of inter-, trans-and multidisciplinarity on modern taxonomy of sciences. *Current Science* 106:685.

Schacht R & Kramer KL. 2019. Are we monogamous? A review of the evolution of pair-bonding in humans and its contemporary variation cross-culturally". *Front Ecol Evol*:10.3389/fevo.2019.00230.

Schmitt S. 2004. *Histoire d'une question anatomique: la repetition des parties*. Museum National d'Histoire Naturelle, Paris.

Schueler GF. 2005. *Reasons & purposes - human rationality and the teleological explanation of actions*. Clarendon Press, Oxford.

Scott JC. 2017. *Against the grain - a deep history of the earliest states*. Yale University Press, New Haven.

Shanahan T. 2017. Selfish genes and lucky breaks: Richard Dawkins' and Stephen Jay Gould's divergent Darwinian agendas. In *The Darwinian tradition in context: research programs in evolutionary biology* (Delisle RG, ed.). Springer, New York, p. 31-36.

Shanavas TO. 2010. Islamic theory of evolution: the missing link between Darwin and the Origin of Species. Brainbow Press. New York.

Sharma A. 1991. Karma and rebirth in Alberuni's India. *Asian Philos* 1:77-91.

Shaw V, Diogo R, Winder I. In press. Hiding in Plain Sight - Revealing the world's oldest anatomical atlas. *Anat Rec.*

Shaw V. 2014. Chong meridian: an ancient Chinese description of the vascular sytem? *Acupunct Med* 32: 10.1136.

Shehata M. 2003. The ear, nose and throat in Islamic medicine. *JISHIM* 1:2-5.

Shermer M. 2004. *The science of good and evil - why people cheat, gossip, care, share, and follow the golden rule.* Times Books, New York.

Shermer M. 2011. *The believing brain - from ghosts and gods to politics and conspiracies, how we construct beliefs and reinforce them as truths.* Times Books, New York.

Shermer M. 2015. *The moral arc - how science makes us better people.* St. Martin's Griffin, New York.

Shoja MM, Tubbs RS. 2007. The history of anatomy in Persia. *J Anat* 210, 359-378.

Shubin N. 2008. *Your inner fish: a journey into the 3.5-billion-year history of the human body.* Vintage, New york.

Siddiqi AH. 1995. Muslim geographic thought and the influence of Greek philosophy. *GeoJournal* 37:9-15.

Singer C. 1997. *A short history of science to the nineteenth century,* 2nd ed. Dover Publications, New York.

Singer CJ. 1941. *A short history of science to the nineteenth century.* Clarendon Press Oxford, Oxford.

Singer CJ. 1957. *A short history of anatomy and physiology from the Greeks to Harvey.* Dover Publications, New York

Singer. CJ. 1959. *History of biology to about the year 1900.* Abelard-Shuman, London.

Singh U. 2008. *A history of ancient and early medieval India: from the Stone Age to the 12th century.* Pearson Education, New York.

Skinner BF. 1948. 'Superstition' in the pigeon. *J Exp Psychol* 38:168-172.

Smith EA, Hill K, Marlowe F, *et al.* 2010. Wealth transmission and inequality among hunter-gatherers. *Curr Anthropol* 51:19-34.

Smith RJ. 2016. Freud and evolutionary anthropology's first just-so story. *Evol Anthropol* 25:50–53.

Sommer M. 2015. *Evolutionäre Anthropologie zur Einführung*. Junius, Hamburg.

Sommer V, Vasey PL, eds. 2006. *Homosexual behavior in animals - an evolutionary perspective*. Cambridge University Press, Cambridge.

Sorenson J. 2009. *Ape*. Reaktion Books Ltd, London.

Souayah N, Greenstein JI. 2005. Insights into neurologic localization by Rhazes, a medieval Islamic physician. *Neurol* 65:125-128.

Sproul B. 1991. *Primal myths - creation myths around the world*. HarperCollins, New York.

Standring S. 2016. A brief history of topographical anatomy. *J Anat* 229:32-62.

Starr SF. 2009. Rediscovering central Asia. *Wilson Quart* 33:33-43.

Stevens B. 2016. *Nihilism - a philosophy based on nothingness and eternity*. Manticore Press, New York.

Stoltzfus A. 2017. Why we don't want another "synthesis". *Biology Direct* 12:23.

Stroumsa S. 2012. Maimonides in his world: portrait of a Mediterranean thinker. Princeton University Press, Princeton.

Sultan SE. 2016. *Organisms & environment - ecological development, niche construction, and adaptation*. Oxford University Press, Oxford.

Suzman J. 2017. *Affluence without abundance - the disappearing world of the Bushmen*. Bloomsbury, New York.

Swan L, Gordon R, Seckbach J, eds. 2012. *Origin(s) of design in nature - a fresh, interdisciplinary look at how design emerges in complex systems, especially life*. Springer, New York.

Syed IB. 2002. *Islamic Medicine: 1000 years ahead of its times*. JISHIM.

Taleb NN. 2010. *The black swan - the impact of the highly improbable, 2nd Edition*. Trader House Trade Paperback, New York.

Thagard P. 2010. *The brain and the meaning of life*. Princeton University Press, Princeton.

Thomas K. 1983. *Man and the natural world - changing attitudes in England 1500-1800*. Oxford University Press, Oxford.

Tibi S. 2006. Al-Razi and Islamic medicine in the 9th century. *J R Soc Med* 99:206-207.

Todes DP. 1989. *Darwin without Malthus*. Oxford University Press, Oxford.

Trüper H, Chakrabarty D, Subrahmanyam S, eds. 2015. *Historical teleologies in the modern world*. Bloomsbury, New York.

Tubbs RS, Shoja MM, Loukas M, Oakes WJ. 2007. Abubakr Muhammad ibn Zakaria Razi, Rhazes (865-925 AD). *Child's Nerv Syst* 23:1225-1226.

Tulp NP. 1641. *Observationes Medicae*. Vivie, Leiden.

Turner DD. 2017. Paleobiology's uneasy relationship with the Darwinian tradition: stasis as data. In *The Darwinian tradition in context: research programs in evolutionary biology* (Delisle RG, ed.). Springer, New York, p. 333-352.

Turner JS. 2000. *The extended organism - the physiology of animal-built structures*. Harvard University Press, Cambridge.

Turner JS. 2007. *The tinkerer's accomplice: how design emerges from life itself*. Harvard University Press, Cambridge.

Turner JS. 2013. Biology's second law: homeostasis, purpose and desire. In *Beyond Mechanism: Putting Life Back into Biology* (Henning BG, Scarfe AC, eds). Lexington Books, Lexington, p. 183-204.

Turner JS. 2016. Homeostasis and the physiological dimension of niche construction theory in ecology and evolution. *Evol Ecol* 30:203-219.

Tuttle RH, ed. 1975. *Primate Functional Morphology and Evolution*. Aldine, Chicago.

Tyson E. 1699. *Orang-Outang sive Homo sylvestris, or the anatomy of a pygmie compared to that of a monkey, an ape and a man*. T. Bennet, London.

Unal N, Elcioglu O. 2009. Anatomy of the eye from the view of ibn Al-Haitham (965-1039). The founder of modern optics. Saudi Med J 30:323-328.

UNESCO. 1950. U.N.E.S.C.O. on Race. *Man* 50:138-139.

UNESCO. 1951. U.N.E.S.C.O.'s New Statement on Race. *Man* 51:154-155.

UNESCO. 1952. U.N.E.S.C.O.'s New Statement on Race. *Man* 52:9.

Vaas R. 2009. Gods, gains and genes - on the natural origin of religiosity by means of bio-cultural selection. In *The biological evolution of religious mind and behavior* (Voland E, Schiefennhovel W, eds). Springer, New York, p. 25-49.

Van Arsdale A. 2017. Human evolution as a theoretical model for an extended evolutionary synthesis. In *The Darwinian tradition in context: research programs in evolutionary biology* (Delisle RG, ed.). Springer, New York, p. 105-130.

Van Schaik C, Michel K. 2016. *The good book of human nature - an evolutionary reading of the bible*. Basic Books, New York City.

Van Wyhe J, Kjaergaard PC. 2015. Going the whole orang: Darwin, Wallace and the natural history of orangutans. *Studies Hist Philos Sci C* 51:53-63.

Varki A. 2009. Human uniqueness and the denial of death. *Nature* 460: 684.

Veracini C, Teixeira DM. 2016. Perception and description of New World non-human primates in the travel literature of the fifteenth and sixteenth centuries: a critical review. *Annals Sci* 74:25-65.

Vesalius A. 1543. *De humani corporis fabrica libri septem*. Ex officina Joannis Oporini, Basel.

Vinicius M. 2012. *Modular evolution: how natural selection produces biological complexity*. Cambridge University Press, Cambridge.

Voland E, Schiefennhovel W, eds. 2009. *The biological evolution of religious mind and behavior*. Springer, New York.

Voland E. 2009. Evaluating the evolutionary status of religiosity and religiousness. In *The biological evolution of religious mind and behavior* (Voland E, Schiefennhovel W, eds). Springer, New York, p. 9-24.

Wagner A. 2014. *Arrival of the fittest: solving evolution's greatest puzzle*. Oneworld publications, London.

Wallace AR. 1853. *Palm trees of the Amazon and their uses*. Biodiversity Heritage Library, London.

Wallace AR. 1869. *The Malay Archipelago*. Harper, London.

Wallace AR. 1870. *Contributions to the theory of natural selection, 2nd ed.* Macmillan and Company, London.

Wallace AR. 1876. *The geographical distribution of Animals*. Harper and brothers, London.

Wallace AR. 1878. *Tropical nature, and other essays*. Macmillan and Company, London.

Wallace AR. 1881. *Island life*. Harper and brothers, London.

Wallace AR. 1889. *Darwinism: an exposition of the theory of natural selection, with some of its applications*. Macmillan and Company, London.

Wallace AR. 1889. *Travels on the Amazon and Rio Negro*. Ward, Lock, Bowden & Co, London.

Wallace AR. 1903. *Man's place in the universe* (Gutenberg). Chapman & Hall, London.

Wallace AR. 1905. My life. Chapman & Hall, London.

Wang Y, Liu H, Sun Z. 2017. *Lamarck rises from his grave: parental environment-induced epigenetic inheritance in model organisms and humans.* Biol Rev 92:2084-2111.

Washington HA. 2006. *Medical apartheid: the dark history of medical experimentation on black Americans from colonial times to the present.* Anchor Books, New York.

Weber BH, Depew DJ, eds. 2003. *Evolution and learning: the Baldwin effect reconsidered.* MIT Press, Cambridge.

West JB. 2008. Ibn Al-Nafis, the pulmonary circulation, and the Islamic Golden Age. *J App Physiol* 105:1877-1880.

West-Eberhard MJ. 2003. *Developmental plasticity and evolution.* Oxford University Press, Oxford.

West-Eberhard MJ. 2007. Dancing with DNA and flirting with the ghost of Lamarck. *Biol & Philos* 22:439-451.

West-Eberhard MJ. 2014. *Darwin's forgotten idea: the social essence of sexual selection. Neurosci Biobehav Rev* 46:501-508.

Westneat DF, Fox CW, eds. 2010. *Evolutionary behavioral ecology.* Oxford University Press, New York.

Wetherington RK. 2011. *Readings in the history of evolutionary theory.* Oxford University Press, Oxford.

White C. 1799. *An account of the regular gradation in Man.* C. Dilly, London.

Wilczynski JZ. 1959. *On the presumed Darwinism of Alberuni eight hundred years before Darwin.* Isis 50:459-466.

Wilson EO. 2014 *The meaning of human existence.* W. W. Norton & Company, New York.

Wray GA, Hoekstra HE, Futuyma DJ, *et al.* 2014. Does evolutionary theory need a rethink? No, all is well. *Nature* 514:161-164.

Wulf A. 2015. *The invention of nature - the adventures of Alexander von Humboldt, the lost hero of science.* Alfred A. Knopf, New York.

Yarmohammadi H, Dalfardi B, Ghanizadeh A. 2013a. Joveini (Al-Akhawayni) and the early knowledge on circle of Willis. *Int J Cardiol* 168:4482-4482.

Yarmohammadi H, Dalfardi B, Rezaian J, Ghanizadeh A. 2013b. Al-Akhawayni's description of pulmonary circulation. *Int J Cardiol* 168:1819-1821.

Zargaran, A., Zarshenas, M. M., Ahmadi, S. A. and Vessal, K. 2013. Haly Abbas (949–982 AD). *J Neurol* 260:2196-2197.

Zarshenas MM, Zargaran A., Mehdizadeh A, Mohagheghzadeh A. 2016. Mansur ibn Ilyas (1380–1422 AD): A Persian anatomist and his book of anatomy, Tashrih-i Mansuri. *J Med Biogr* 24:67-21.

Ziaee A. 2014. Persian illustrated anatomy from Timurid Iran. *IOSR-JHSS* 19:49-52.

Zirkle C. 1941. Natural selection before the "Origin of Species". *Proc Am Philos Soc* 84:71-123.

FIGURE CREDITS

Fig. 0 (cover). Image freely available, and copyright free, from https://en.wikipedia.org/wiki/Al-Jahiz#/media/File:Al-Jahiz.jpg

Fig. 1.1. Image freely available, and copyright free, from https://en.wikipedia.org/wiki/Galen#/media/File:Claudius_Galenus_(1906)_-_Veloso_Salgado.png

Fig. 1.2. Image freely available, and copyright free, from https://en.wikipedia.org/wiki/Avicenna#/media/File:Avicenna_Portrait_on_Silver_Vase_-_Museum_at_BuAli_Sina_(Avicenna)_Mausoleum_-_Hamadan_-_Western_Iran_(7423560860).jpg

Fig. 1.3. Image freely available, and copyright free, from https://en.wikipedia.org/wiki/Andreas_Vesalius#/media/File:De_humani_corporis_fabrica_(27).jpg

Fig. 1.4. Image freely available from https://upload.wikimedia.org/wikipedia/commons/9/95/Portrait_of_A.R._Wallace._Wellcome_L0000531.jpg.

Fig. 1.5. Image freely available from https://upload.wikimedia.org/wikipedia/commons/f/f8/Portrait_of_Charles_Darwin_Wellcome_M0011462.jpg.

Fig. 1.6. Image freely available and adapted from https://upload.wikimedia.org/wikipedia/commons/5/56/Ushuaia-yamana7.jpg.

Fig. 1.7. Image freely available and adapted from http://www.chileparaninos.gob.cl/639/w3-article-321019.html.

Fig. 1.8. Figure freely available and adapted from https://allthatsinteresting.com/hypatia-of-alexandria.

Fig. 1.9. Due to its antiquity, this figure has no copyright; adapted from https://www.dreamstime.com/ascent-life-vintage-engraving-engraved-illustration-earth-man-image162956266.

Fig. 1.10. Joseph Wright of Derby, figure freely available from https://upload.wikimedia.org/wikipedia/commons/0/0a/Erasmus_Darwin%2C_after_Joseph_Wright.jpg.

Fig. 1.11. Text and figure freely available, and copyright free, from https://www.visualcapitalist.com/how-people-spend-their-time-globally/
Fig. 1.12. Modified from Kelly 2013.
Fig. 2.1. Image freely available, and copyright free, from https://commons.wikimedia.org/wiki/File:Macaca_sylvanus.Mother_and_baby.jpg
Fig. 2.2. Image modified from our Alghamdi et al. 2017 paper.
Fig. 2.3. Image modified from our Alghamdi et al. 2017 paper.
Fig. 2.4. Image modified from our Alghamdi et al. 2017 paper.
Fig. 2.5. Image modified from our Alghamdi et al. 2017 paper.
Fig. 2.6. Image modified from our Alghamdi et al. 2017 paper.
Fig. 2.7. Image modified from our Alghamdi et al. 2017 paper.
Fig. 2.8. Image modified from our Alghamdi et al. 2017 paper.
Fig. 2.9. Image modified from our Alghamdi et al. 2017 paper.
Fig. 3.1. Image freely available, and copyright free, from https://www.pewforum.org/2009/02/04/religious-differences-on-the-question-of-evolution/
Fig. 3.2. Image modified from our Malik et al. 2017 paper.
Fig. 3.3. Image modified from our Malik et al. 2017 paper.
Fig. 3.4. Image freely available, and copyright free, from https://en.wikipedia.org/wiki/Rumi#/media/File:Meeting_of_Jalal_al-Din_Rumi_and_Molla_Shams_al-Din.jpg
Fig. 3.5. Image freely available, and copyright free, from https://en.wikipedia.org/wiki/Al-Jahiz#/media/File:Al-Jahiz.jpg
Fig. 4.1. Image modified from our Campbell et al. 2021 paper.
Fig. 4.2. Image freely available, and copyright free, from https://medium.com/paperkin/where-is-evolution-taking-the-human-race-6dda-f7eaddba
Fig. 4.3. Image freely available, and copyright free, from https://twitter.com/archaeologyEAA/status/1372105138236706818/photo/2

INDEX

A

Aberrhos 116
acculturation 9, 13
active hunting 26
adaptationism 21
adaptationists 23
adaptationist way of thinking 21
African apes 139
Al-Akhawayni Bukhari 54
Al-Baghdadi 48, 99, 100, 101, 103, 110, 111
Al-Beruni 122, 124, 132, 133, 134, 140
Alfred Russel Wallace 4
Al-Jahiz xi, 122, 124, 125, 126, 127, 142
Al-Qaida 113
Al-Razi 48, 49, 50, 51, 52, 53, 55, 56, 58, 59, 61, 72, 73, 75, 76, 98, 106, 110
American slavery 29
anatomical sciences 35, 56, 110
Ancient China 4, 112, 147
Ancient Egypt 4
Ancient Greece 35, 143, 147, 151
Ancient Greek 17, 18, 133, 139, 143
animal behaviour 22
animal homosexual behaviour 22
Animism 28
Antwerp zoo 29
ape-like 133
Arabs 116, 148
argument from design 19
Aristotle 3, 16, 17, 18, 71, 98, 104, 106, 116, 124, 130, 139, 143, 147
artificial selection 14, 133
Asa Gray 21
assimilationists 29, 30
astronomy 6, 14, 15, 149
Averroes xiv, 6, 95, 116, 134
Avicenna 2, 35, 48, 73, 74, 116

B

Barbary ape 35
Beagle Channel 11
Beagle voyage 10, 114
behavioral ecology 21
behavioural 'abnormalities' 22
Belgian Congo 28
biased narratives 2, 13, 114
Bible 135
biogenesis 135
biological fitness 126
biology classes 118
biology students 117, 118
body systems 45
Boko Haram 114
Book of Animals 124
BouJaoude 117, 118

C

capitalism 31
capitalist 145
capitalistic ideas 14
cave painters 152
chain of being 18, 151
chain of life 151
Charles 5, 7, 12, 19, 20, 142
Christianity 46
Christopher Columbus 4
circular adaptationist reasoning 23
college students 117
colonialism 28, 29, 113, 114, 146

colonial propaganda 28
communism 31
comparative anatomy 116, 152
competition 14
Confederacy 29
Constantinus Africanus 56
Copernicus xiii, 6
Cotton Mather 30
Covid-19 pandemic 16, 30
crabs 18
Crusades 148

D

Daniel Defoe 135
Darwin i, iii, ix, xi, xii, xiii, xiv, 4, 5, 6, 7, 8, 9, 11, 12, 13, 14, 15, 16, 17, 18, 21, 23, 25, 114, 115, 121, 122, 124, 126, 128, 132, 133, 134, 136, 142, 143, 147,
David Attenborough 16
Da Vinci 3
Descartes xiii, 6
discrimination 29, 30, 129, 152

E

Egypt 15, 100, 118
Egyptians xi, 35, 147
elephants 130
Empedocles 139
empiricism 146
Enlightenment 148
environmental conditions 20, 126, 127, 140
epigenetics 31
Erasmus Darwin 19, 121, 142

Ernst Mayr 116
eschatology of progress 145
esoteric society 129
Euclid 8, 16, 94
eugenics 146
European Renaissance 2, 3, 110
evolutionary adaptation 21, 23
evolutionary medicine 21
evolutionary psychology 21, 23

F

father of human anatomy 35
first artificial Earth satellite 126
food gatherers 152
Francisco Ayala 121
Francis Darwin 7

Freud 27, 28
Freud's 'psychoanalytic theories' 28
Fuegians 9, 10, 11, 12
functional morphology 45

G

Galen xii, 1, 2, 3, 35, 36, 46, 47, 48, 51, 52, 53, 55, 56, 58, 59, 61,

63, 71, 72, 75, 76, 88, 92, 94, 96, 97, 98, 100, 101, 102, 103, 104, 105, 106, 107, 110, 112, 115
genes 25, 31
genetic and epigenetic factors 140, 142
genetics 31, 32
genocide 146
George Edwards 22
Golden Islamic Age 47, 114, 116, 132
great chain of being 130
Greco-Roman culture of the dead body 35
Greek philosophy 148

H

Haldane 136
Hazda 26
heterosexual 23
Hippocrates 55, 56, 57, 58, 72, 106
Homo antecessor 151
homosexuality 22
Hrdy 26
human fossils 149
hunter-gatherer societies 26
hunters 24, 25, 27
Huxley 128
Hypatia 14, 15

I

Ibn Abbas 56, 57, 58, 59, 60, 61, 62, 63, 71, 72, 73, 75, 76
ibn Al-Haytham 48, 93
Ibn Al-Nafis 101, 104, 105
Ibn Khaldun 122, 141, 142
Ibn Miskawayh 122, 127, 129
Ibn Rushd 95, 97
Ibn Sina 73, 74, 75, 76, 88, 89, 90, 91, 92, 104
ibn Tufail 134
ibn Tufayl 115, 134, 135, 136, 137
ideological religious arguments 117
immunology, 30
imperialism 113, 114, 146, 148
imperial regimes 148
industrial 145
inoculation 30
intelligent design 16, 143
intelligent designer 16
intentionality in nature 18
ISIL 113
ISIS 113
Islam 46, 48, 113, 114, 120, 148, 155
Islamic empires 148
Islamic Golden Age xiv, 2, 3, 6, 110, 139, 142
Islamic philosophy 135, 139
Islamic State of Iraq and Syria 113
Islamic theology and jurisprudence 148

J

Jefferson Davis 29
Jemmy 10, 12
John William Draper 115
José María Bermúdez de Castro 151

K

Karl Marx 14
Keosian 136

L

ladder of life 17, 18, 140, 141, 143
ladder-of-life 122, 123, 130, 131, 133, 138, 139, 143, 151
Lamarck 121, 142

Lebanon 117, 118, 119, 120
liberalism 14
Lucy 149, 150

Lysenko 31, 32
Lysenkoism 32

M

Macaca sylvanus 35
Malthusian struggle for existence 14
Mansur ibn Ilyas 48, 105, 106
mariner's astrolabe 4
marriage 23, 24
Marxism 31
material sciences 148, 149
mathematics 6, 7, 14, 93, 116, 129, 147, 148, 149
Mayan territories 4
medicine xi, 2, 6, 35, 47, 50, 54, 74, 94, 95, 100, 116, 139, 148, 149
misogynistic biases 25
Modernity 145
modern vaccination 30
monkeys 26, 36, 58, 63, 77, 111, 128, 132, 141, 143
Mother Nature xiv, 6, 15, 19, 21, 124, 126, 133, 134, 138
Musa al-Gharbi 145
Muzaffar Iqbal 114

N

Nabataeans 127
natural selection xii, xiv, 4, 5, 6, 14, 20, 121, 124, 126, 128, 132, 133, 140
Natural Theologians 19

Natural Theology 8
Newton xiii, 6
Nidhami Arudi 122, 124, 137
Nikolai Vavilov 31, 32
Nunamiut 26

O

objectivity 146
Onesimus 30
oppression 29, 152

orangutans 26, 139
'orientalist' scholarly enterprise 149
origin of life 135, 136

P

Paley 8, 16, 19
philosophy 1, 6, 15, 16, 18, 124, 129, 137, 139
physical sciences 148
physiology 2, 3, 18, 45, 54, 101, 116
platonic teleology 18

political teleological narratives 31
positivism 146
progress 14, 20, 111, 151, 155
provider theory of marriage 23
purpose 3, 15, 18

Q

Queen Victoria 15
Qur'an 148

Qur'anic exegesis 148
Quranic verses 135

R

rationality 146
religion xiv, 6, 118, 120, 145
religious or anti-evolution perspective 117
Robinson Crusoe 135

Royal Belgian Institute of Natural Sciences 28
Rudyard Kipling 135
Rumi ix, 1, 113, 122, 123

S

same-sex sexual activity 22
same-sex sexual behaviour 22
satyr 139
scala naturae 17, 129
scientific biases xiv, 22, 30, 31, 146
scientific method 1, 4, 9, 136
scientific or religious-reconciliation perspective; 117
scientific racism 27, 149, 152
scientific thinking 136
secular 145
segregationists 29
segregationist thinking 29
selfishness 14

Sierra de Atapuerca 151
Sigmund Freud 27
skin pigmentation 150
smallpox 30
Smithsonian National Museum of Natural History 150
Soviet biology 32
Soviet Union 31, 126
Sputnik 1 126
Stalin 31, 32
Stephen Jay Gould 21
superiority of Christianity and Western civilization 148

T

tafsir 148
Taung Child 149
Teleological narratives 15
teleological view 15, 19
teleology 17, 18, 20, 21
The Ikhwan Al-Safa 122, 129
theology 6, 129

thermoregulation 16
Thomas Robert Malthus 14
Tierra del Fuego 9, 10, 11
Tommaso d'Aquino xiv, 6
traditional gender roles 152
Tusi 115, 122, 124, 139, 140

U

unfit for survival 146
universality 146

urban 145

V

vaccines 30
Vesalius xii, xiii, 3, 46, 47, 48, 97, 110, 111, 112

W

Wallace xi, xii, 4, 5, 18, 121, 124, 126, 128, 133, 136, 142
Western evolutionary biology 142
Western social science 146, 148, 149
William Douglass 31
William Paley 16

Z

Zoarastrianism 28

www.ingramcontent.com/pod-product-compliance
Lightning Source LLC
Chambersburg PA
CBHW020759160426
43192CB00006B/384